ENVIRONMENTAL SCIENCE, ENGINEERING AND TECHNOLOGY

GENETICALLY ENGINEERED CROPS IN AMERICA

ANALYSES, ADOPTION, TRENDS

ENVIRONMENTAL SCIENCE, ENGINEERING AND TECHNOLOGY

Additional books in this series can be found on Nova's website under the Series tab.

Additional e-books in this series can be found on Nova's website under the e-book tab.

ENVIRONMENTAL SCIENCE, ENGINEERING AND TECHNOLOGY

GENETICALLY ENGINEERED CROPS IN AMERICA

ANALYSES, ADOPTION, TRENDS

FREDRICK G. LAWRENCE
EDITOR

nova publishers
New York

Library of Congress Cataloging-in-Publication Data

ISBN: 978-1-63321-225-1

Published by Nova Science Publishers, Inc. † New York

CONTENTS

Preface		**vii**
Chapter 1	Genetically Engineered Crops in the United States *Jorge Fernandez-Cornejo, Seth Wechsler, Mike Livingston and Lorraine Mitchell*	**1**
Chapter 2	The First Decade of Genetically Engineered Crops in the United States *Jorge Fernandez-Cornejo and Margriet Caswell*	**75**
Chapter 3	Private Research and Development for Crop Genetic Improvement *Paul W. Heisey and Keith O. Fuglie*	**115**
Index		**151**

PREFACE

Genetically engineered (GE) varieties with pest management traits became commercially available for major crops in 1996. More than 15 years later, adoption of these varieties by U.S. farmers is widespread and U.S. consumers eat many products derived from GE crops—including cornmeal, oils, and sugars—largely unaware that these products were derived from GE crops. Despite the rapid increase in the adoption of corn, soybean, and cotton GE varieties by U.S. farmers, questions persist regarding their economic and environmental impacts, the evolution of weed resistance, and consumer acceptance. This book examines issues related to three major stakeholders in agricultural biotechnology: GE seed suppliers and technology providers (biotech firms), farmers, and consumers.

Chapter 1 – More than 15 years after their first successful commercial introduction in the United States, genetically engineered (GE) seeds have been widely adopted by U.S. corn, soybean, and cotton farmers. Still, some questions persist regarding the potential benefits and risks of GE crops. The report finds that, although the pace of research and development (measured by the number of USDA-approved field tests) peaked in 2002, other measures show that biotech firms continue to develop new GE seed varieties at a rapid pace. Also, U.S. farmers continue to adopt GE seeds at a robust rate, and seed varieties with multiple (stacked) traits have increased at a very rapid rate. Insecticide use has decreased with the adoption of insect-resistant crops, and herbicide-tolerant crops have enabled the substitution of glyphosate for more toxic and persistent herbicides. However, overreliance on glyphosate and a reduction in the diversity of weed management practices have contributed to the evolution of glyphosate resistance in some weed species.

Chapter 2 – Ten years after the first generation of genetically engineered (GE) varieties became commercially available, adoption of these varieties by

U.S. farmers is widespread for major crops. Driven by farmers' expectations of higher yields, savings in management time, and lower pesticide costs, the adoption of corn, soybean, and cotton GE varieties has increased rapidly. Despite the benefits, however, environmental and consumer concerns may have limited acceptance of GE crops, particularly in Europe. This chapter focuses on GE crops and their adoption in the United States over the past 10 years. It examines the three major stakeholders of agricultural biotechnology and finds that (1) the pace of R&D activity by producers of GE seed (the seed firms and technology providers) has been rapid, (2) farmers have adopted some GE varieties widely and at a rapid rate and benefited from such adoption, and (3) the level of consumer concerns about foods that contain GE ingredients varies by country, with European consumers being most concerned.

Chapter 3 – Seed has been an essential input in crop production since the origins of agriculture, when farmers first began to save grains for replanting and to select seeds for desirable characteristics. The pace of crop genetic improvement accelerated early in the 20th century with the development of applied genetics and associated changes in plant breeding, seed production, and seed marketing. The development of the modern seed industry began about this time. From the mid-1990s to the present, the private-sector seed industry has probably undergone more structural change than any of the other agricultural input industries covered in this study. Technological innovation in the form of modern, DNA-level biotechnology and changes in intellectual property rules have enabled private-sector companies to capture more value from the new seeds they develop. In the late 1990s, research intensity (R&D spending as a percentage of sales) in the seed industry accelerated past intensity in all other agricultural input sectors, and seed remains the most research-intensive sector to date.

In: Genetically Engineered Crops in America ISBN: 978-1-63321-225-1
Editor: Fredrick G. Lawrence

Chapter 1

GENETICALLY ENGINEERED CROPS IN THE UNITED STATES*

Jorge Fernandez-Cornejo, Seth Wechsler, Mike Livingston and Lorraine Mitchell

ABSTRACT

More than 15 years after their first successful commercial introduction in the United States, genetically engineered (GE) seeds have been widely adopted by U.S. corn, soybean, and cotton farmers. Still, some questions persist regarding the potential benefits and risks of GE crops. The report finds that, although the pace of research and development (measured by the number of USDA-approved field tests) peaked in 2002, other measures show that biotech firms continue to develop new GE seed varieties at a rapid pace. Also, U.S. farmers continue to adopt GE seeds at a robust rate, and seed varieties with multiple (stacked) traits have increased at a very rapid rate. Insecticide use has decreased with the adoption of insect-resistant crops, and herbicide-tolerant crops have enabled the substitution of glyphosate for more toxic and persistent herbicides. However, overreliance on glyphosate and a reduction in the diversity of weed management practices have contributed to the evolution of glyphosate resistance in some weed species.

* This is an edited, reformatted and augmented version of a report, ERR-162, issued by the U.S. Department of Agriculture, Economic Research Service, February 2014.

Keywords: Genetically engineered crops, agricultural biotechnology, seed industry, research and development, adoption, crop yields, pesticide use, corn, soybeans, cotton

What Is the Issue?

Genetically engineered (GE) varieties with pest management traits became commercially available for major crops in 1996. More than 15 years later, adoption of these varieties by U.S. farmers is widespread and U.S. consumers eat many products derived from GE crops—including cornmeal, oils, and sugars—largely unaware that these products were derived from GE crops. Despite the rapid increase in the adoption of corn, soybean, and cotton GE varieties by U.S. farmers, questions persist regarding their economic and environmental impacts, the evolution of weed resistance, and consumer acceptance.

What Did the Study Find?

This report examines issues related to three major stakeholders in agricultural biotechnology: GE seed suppliers and technology providers (biotech firms), farmers, and consumers.

GE seed suppliers/technology providers. The number of field releases for testing of GE varieties approved by USDA's Animal and Plant Health Inspection Service (APHIS) is an important measure of research and development (R&D) activities in agricultural biotechnology. The number of releases grew from 4 in 1985 to 1,194 in 2002 and averaged around 800 per year thereafter. However, while the number of releases peaked in 2002, other measures of research and development activity—the number of sites per release and the number of gene constructs (ways that the gene of interest is packaged together with other elements)—have increased very rapidly since 2005. Also, releases of GE varieties with agronomic properties (like drought resistance) jumped from 1,043 in 2005 to 5,190 in 2013.

As of September 2013, about 7,800 releases were approved for GE corn, more than 2,200 for GE soybeans, more than 1,100 for GE cotton, and about 900 for GE potatoes. Releases were approved for GE varieties with herbicide tolerance (6,772 releases), insect resistance (4,809), product quality such as flavor or nutrition (4,896), agronomic properties like drought resistance (5,190), and virus/fungal resistance (2,616). The institutions with the most authorized field releases include Monsanto with 6,782, Pioneer/DuPont with 1,405, Syngenta with 565, and USDA's Agricultural Research Service with 370. As of September 2013, APHIS had received 145 petitions for deregulation (allowing GE seeds to be sold) and had approved 96 petitions: 30 for corn; 15 for cotton; 11 for tomatoes; 12 for soybeans; 8 for rapeseed/canola; 5 for potatoes; 3 for sugarbeets; 2 each for papaya, rice, and squash; and 1 each for alfalfa, plum, rose, tobacco, flax, and chicory.

Farmers. Three crops (corn, cotton, and soybeans) make up the bulk of the acres planted to GE crops. U.S. farmers planted about 169 million acres of these GE crops in 2013, or about half of total land used to grow crops. Herbicide-tolerant (HT) crops have traits that allow them to tolerate more effective herbicides, such as glyphosate, helping adopters control pervasive weeds more effectively. U.S. farmers used HT soybeans on 93 percent of all planted soybean acres in 2013.

HT corn accounted for 85 percent of corn acreage in 2013, and HT cotton constituted 82 percent of cotton acreage. Farmers planted insect-resistant (Bt) cotton to control pests such as tobacco budworm, cotton bollworm, and pink bollworm on 75 percent of U.S. acreage in 2013. Bt corn—which controls the European corn borer, the corn rootworm, and the corn earworm—was planted on 76 percent of corn acres in 2013.

The adoption of Bt crops increases ***yields*** by mitigating yield losses from insects. However, empirical evidence regarding the effect of HT crops on yields is mixed. Generally, stacked seeds (seeds with more than one GE trait) tend to have higher yields than conventional seeds or than seeds with only one GE trait. GE corn with stacked traits grew from 1 percent of corn acres in 2000 to 71 percent in 2013. Stacked seed varieties also accounted for 67 percent of cotton acres in 2013.

Planting Bt cotton and Bt corn seed is associated with higher ***net returns*** when pest pressure is high. The extent to which HT adoption affects net returns is mixed and depends primarily on how much weed control costs are reduced and seed costs are increased. HT soybean adoption is associated with an increase in total household income because HT soybeans require less management and enable farmers to generate income via off-farm activities or by expanding their operations.

Farmers generally use less ***insecticide*** when they plant Bt corn and Bt cotton. Corn insecticide use by both GE seed adopters and nonadopters has decreased—only 9 percent of all U.S. corn farmers used insecticides in 2010. Insecticide use on corn farms declined from 0.21 pound per planted acre in 1995 to 0.02 pound in 2010. This is consistent with the steady decline in European corn borer populations over the last decade that has been shown to be a direct result of Bt adoption. The establishment of minimum refuge requirements (planting sufficient acres of the non-Bt crop near the Bt crop) has helped delay the evolution of Bt resistance. However, there are some indications that insect resistance is developing to some Bt traits in some areas.

The adoption of HT crops has enabled farmers to substitute glyphosate for more toxic and persistent ***herbicides***. However, an overreliance on glyphosate and a reduction in the diversity of weed management practices adopted by crop producers have contributed to the evolution of glyphosate resistance in 14 weed species and biotypes in the United States. Best management practices (BMPs) to control weeds may help delay the evolution of resistance and sustain the efficacy of HT crops. BMPs include applying multiple herbicides with different modes of action, rotating crops, planting weed-free seed, scouting fields routinely, cleaning equipment to reduce the transmission of weeds to other fields, and maintaining field borders.

The ***price*** of GE soybean and corn seeds grew by about 50 percent in real terms (adjusted for inflation) between 2001 and 2010. The price of GE cotton seed grew even faster. The yield advantage of Bt corn and Bt cotton over conventional seed has become larger in recent years as new Bt traits have been incorporated and stacked traits have become available. Planting Bt cotton and Bt corn continues to be more profitable, as measured by net returns, than planting conventional seeds.

Consumers. Consumer acceptance of foods with GE ingredients varies with product characteristics, geography, and the information that consumers are exposed to. Most studies in industrialized nations find that consumers are willing to pay a premium for foods that don't contain GE ingredients. However, studies in developing countries yield more mixed results. Some studies, including some with a focus on GE ingredients with positive enhancements (such as nutrition), find consumers to be willing to try GE foods and even to pay a premium for them, while others find a willingness to pay a premium for non-GE foods. Most studies have shown that willingness-to-pay for non-GE foods is higher in the EU, where some retailers have policies limiting the use of GE ingredients. Non-GE foods are available in the United States, but there is evidence that such foods represent a small share of retail food markets.

How Was the Study Conducted?

This report updates the ERS report titled *The First Decade of Genetically Engineered Crops in the United States*. To consider biotech seed firms, we use information from the literature and analyze USDA data on field testing approvals by APHIS for new GE varieties. To study farmers' use of GE crops, we analyze USDA farm surveys, particularly the Agricultural Resource Management Survey (ARMS), and summarize the literature. To understand consumers' perspectives, we summarize surveys of consumers' attitudes from the literature.

Introduction

Genetic engineering is a key component of modern agricultural biotechnology.[1] The first genetically engineered (GE) plant, a tomato, was developed in 1982 (USDA/ARS, 2012). By 1985, the USDA had approved four releases of GE organisms for field testing. Commercial use of major GE crops began in 1996.[2]

Genetically engineered crop traits have been classified into one of three generations (FernandezCornejo, 2004). The first generation features enhanced input traits such as herbicide tolerance, resistance to insects, and resistance to environmental stress (like drought). The second features value-added output traits such as nutrient-enhanced seeds for feed. The third generation of GE crops would include traits to allow production of pharmaceuticals and products beyond traditional food and fiber.

While the first GE crop approved by USDA's Animal and Plant Health Inspection Service (APHIS) and commercialized in 1994 was a crop with a strictly second-generation trait (FlavrSavr tomato), most GE crops planted in the United States have first-generation traits. All three generations of GE crop traits are in various stages of research and development.[3]

Most U.S. acres planted to GE crops have traits that provide herbicide tolerance (HT) and/or insect resistance. These seeds became commercially available in 1996. HT crops are able to tolerate certain highly effective herbicides, such as glyphosate, allowing adopters of these varieties to control pervasive weeds more effectively. Commercially available HT crops include soybeans, corn, cotton, canola, sugarbeets, and alfalfa. Insect-resistant or Bt crops contain a gene from the soil bacterium *Bacillus thuringiensis* (Bt) that produces a protein which is toxic to certain insects, protecting the plant over its entire life (Fernandez-Cornejo and McBride, 2002). Commercially available Bt crops include corn and cotton.

More than 15 years after commercial introduction, adoption of first-generation GE crop varieties by U.S. farmers has reached about 90 percent of the planted acres of corn, soybeans, and cotton. U.S. consumers eat many products derived from these crops—including cornmeal, oils, and sugars—largely unaware of their GE origins. Despite the rapid increase in adoption rates for GE corn, soybean, and cotton varieties by U.S. farmers, some continue to raise questions regarding the potential benefits and risks of GE crops.

This report updates ERS' 2006 report, *The First Decade of Genetically Engineered Crops in the United States.* As in the previous report, this report examines the three major stakeholders of agricultural biotechnology: GE seed suppliers and technology providers (biotech firms), farmers, and consumers.

FROM THE LABORATORY TO THE FIELD

Over the last century, private research and development (R&D) expenditures in the seed industry have increased rapidly both in absolute terms and relative to public expenditures, altering the focus of R&D and of the crops studied (Fernandez-Cornejo, 2004). Over the past two decades, technological innovation in the form of modern biotechnology and changes in property rights have enabled private-sector firms to capture more value from the seeds that they develop, and seed remains the most research-intensive of the agricultural input sectors to date (Heisey and Fuglie, 2012).

While the rapid commercial success of GE varieties is the fulfillment of R&D efforts, earlier benchmarks include the number of releases for field testing of GE plant varieties approved by APHIS as well as the determination of nonregulated status (see box, "Regulatory Oversight").[4] Field testing is a critical part of seed development (Fernandez-Cornejo and Caswell, 2006).

Field Releases

The number of field release permits and notifications issued by APHIS for GE organisms (mostly plant varieties) grew from 4 in 1985 to 1,194 in 2002 and then averaged around 800 per year (figure 1). The cumulative number (beginning in 1985 and ending in September 2013) of releases for field testing increased from 10,700 in 2005 to more than 17,000 in 2013. Field releases approved for corn increased from close to 5,000 in 2005 to 7,800 in 2013. Approved releases for GE varieties with herbicide tolerance traits increased from 3,587 in 2005 to 6,772 in 2013, insect resistance from 3,141 to 4,909, and product quality such as flavor or nutrition from 2,314 to 4,896.

REGULATORY OVERSIGHT

Before commercial introduction, genetically engineered (GE) crops must conform to standards set by State and Federal statutes (Fernandez-Cornejo and Caswell, 2006; USDA/APHIS, 2013). Under the Coordinated Framework for the Regulation of Biotechnology, Federal oversight is shared by the U.S. Department of Agriculture (USDA), the U.S. Environmental Protection Agency (EPA), and the U.S. Food and Drug Administration (FDA).

USDA's Animal and Plant Health Inspection Service (APHIS) plays a central role in regulating field testing of agricultural biotechnology products. Through either a notification or permit procedure, such products—which include certain genetically engineered plants, microorganisms, and invertebrates—are considered "regulated articles." APHIS issues authorizations for field releases of those GE organisms (mostly GE plants) that are categorized as "regulated articles" under its regulations, to allow technology providers to pursue field testing. GE plants that meet six specific criteria described in the regulations undergo an administratively streamlined process, known as a *notification*. Under a *notification*, applicants provide information on the nature of the plant and introduced genes, descriptions of genetic modifications, size of the introduction, and origin and destinations for movement or the location of a field test. For GE plants that do not meet the criteria for a *notification*, an APHIS *permit* is required. This process involves a more comprehensive review. In addition to the data required for notification, permit applicants must describe how they will perform the test, including specific measures to reduce the risk of harm to other plants, so the tested organisms remain confined and do not persist after completion of the field test.

After years of field tests, an applicant may petition APHIS for a determination of nonregulated status in order to facilitate commercialization of the product.

If, after extensive review, APHIS determines that the GE organism is unlikely to pose a plant pest risk, the organism is issued a "determination of nonregulated status." At this point, the organism is no longer considered a regulated article and can be moved and planted without APHIS oversight under the biotechnology regulations (USDA/APHIS, 2012).

If a plant is engineered to produce a substance that "prevents, destroys, repels, or mitigates a pest," it is considered a pesticide and is subject to regulation by EPA (*Federal Register*, November 23, 1994). FDA regulates all food applications of crops, including those crops that are developed through the use of biotechnology, to ensure that foods derived from new plant varieties are safe to eat. A more complete description of the regulations of GE products may be found in USEPA, 2003; Belson, 2000; and USDA/APHIS, 2013). Though the current regulatory system is considered to be effective, USDA, EPA, and FDA update regulations as needed to address new trends and issues of the future.

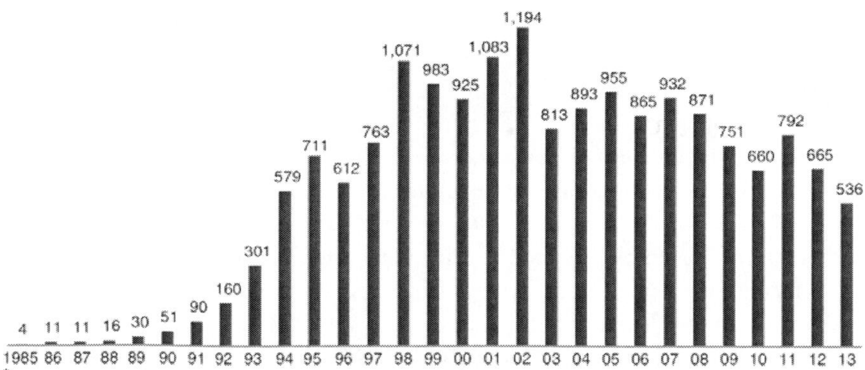

*As of September 24, 2013.

Authorizations for field releases of GE organisms (mostly plant varieties) are issued by USDA's Animal and Plant Health Inspection Service (APHIS) to allow technology providers to pursue field testing.

Source: Information Systems for Biotechnology (ISB, 2013).

Figure 1. Number of releases of genetically engineered (GE) organisms varieties approved by APHIS, 1985-2013* (Includes permits and notifications).

However, these numbers do not fully indicate the amount of R&D activity. A permit or notification can include many release sites and authorize many different gene constructs (ways that the gene of interest is packaged with other elements, like promoters that allow gene expression) to be tested at each site.[5] Thus, while the number of APHIS notifications and permits peaked in 2002, a more comprehensive measure of the amount of R&D activity in agricultural biotechnology—the number of authorized sites and authorized constructs—has increased very rapidly since 2005. For example, while the number of releases authorized in fiscal year (FY) 2012 was lower than in FY2005, the number of authorized sites in FY2012 almost doubled those in FY2005 and the number of constructs increased more than 150-fold (table 1).[6]

Table 1. Number of releases, sites, and constructs authorized by APHIS for evaluation

	Releases	Authorized sites	Authorized constructs
FY2012	767	9,133	469,202
FY2011	967	10,128	395,501
FY2010	754	6,626	297,422
FY2009	846	6,751	217,502
FY2008	948	7,744	125,365
FY2007	1,066	3,623	63,217
FY2006	974	4,327	18,532
FY2005	1011	4,939	3,042
FY2004	997	4,523	2,851
FY2003	824	2,910	2,650
FY2002	1,226	5,111	3,234
FY2001	1,190	5,831	3,208
FY2000	1,002	3,836	3,126
FY1999	1,068	4,134	3,502
FY1998	1,151	4,781	3,830
FY1997	782	3,427	2,650
FY1996	653	2,745	2,305
FY1995	734	3,690	2,666
FY1994	569	1,669	1,926
FY1993	341	455	870
FY1992	164	121	427
FY1991	90	10	226
FY1990	46	14	142
FY1989	32	12	74

A gene construct is the name used for a functional unit necessary for the transfer or the expression of a gene of interest (http://www.gmo-safety.eu/glossary/667.gene-construct.html). Typically, a construct comprises the gene or genes of interest, a marker gene (to facilitate detection inside the plant), and appropriate control sequences as a single package (Food and Agriculture Organization, 2001).

A construct is a piece of DNA which functions as the vehicle or vector carrying the target gene into the recipient organism. It has several different regions.

Source: Unpublished USDA Animal and Plant Health Inspection Service database.

Most field releases have involved major crops, particularly corn, which had about 7,800 field releases approved as of September 2013. More than 2,200 field releases were approved for GE soybeans, more than 1,100 for GE cotton, and about 900 for GE potatoes (figure 2). Releases approved between 1985 and September 2013 included GE varieties with herbicide tolerance (6,772), insect resistance (4,809), product quality such as flavor or nutrition (4,896), agronomic properties (like drought resistance) (5190), and virus/ fungal resistance (2,616) (figure 3). A notable change in R&D activities between 2005 and 2013, as measured by the field releases of GE varieties, is the five-fold jump in releases of GE varieties with agronomic properties (like drought resistance) from 1,043 in 2005 to 5,190 in 2013 (figure 3).

The top release permit-holding institutions include Monsanto (6,782 permits/notifications held), Pioneer/DuPont (1,405), Syngenta (565), and USDA/ARS (370) (figure 4).

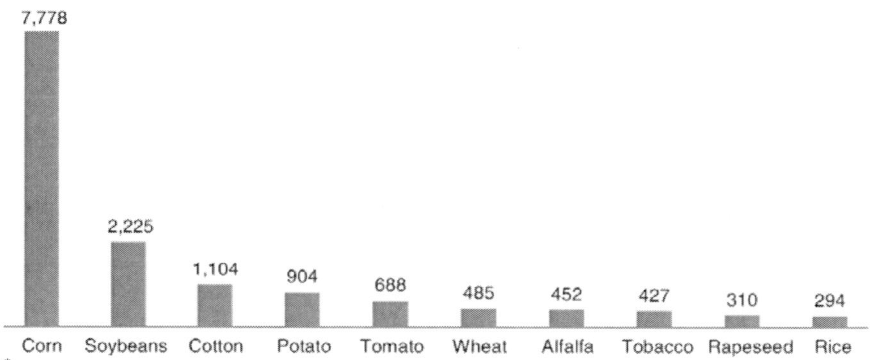

*As of September 24, 2013.

Authorizations for field releases of GE plant varieties are issued by USDA's Animal and Plant Health Inspection Service (APHIS) to allow technology providers to pursue field testing.

Source: Information Systems for Biotechnology (ISB, 2013).

Figure 2. Number of releases approved by APHIS: Top 10 crops (includes permits and notifications).*

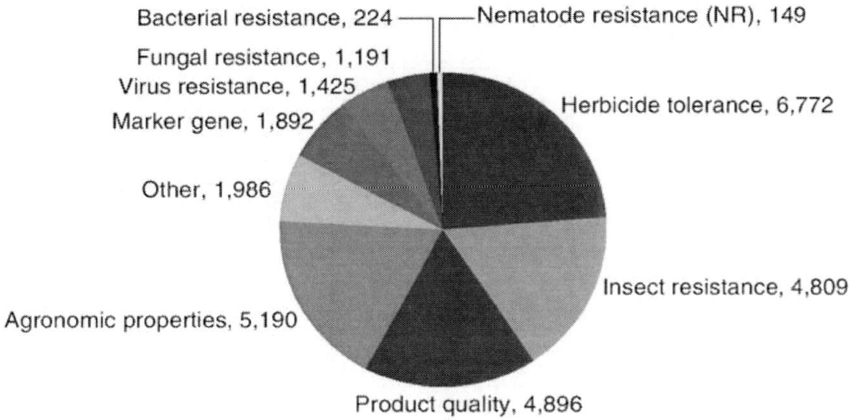

*As of September 24, 2013.

Authorizations for field releases of GE plant varieties are issued by USDA's Animal and Plant Health Inspection Service (APHIS) to allow technology providers to pursue field testing. Counts refers to the actual number of approved release locations per phenotype category. http://www.aphis.usda.gov/biotechnology/status.shtml

Source: Information Systems for Biotechnology (ISB, 2013).

Figure 3. Number of releases approved by APHIS by GE trait (includes permits and notifications).*

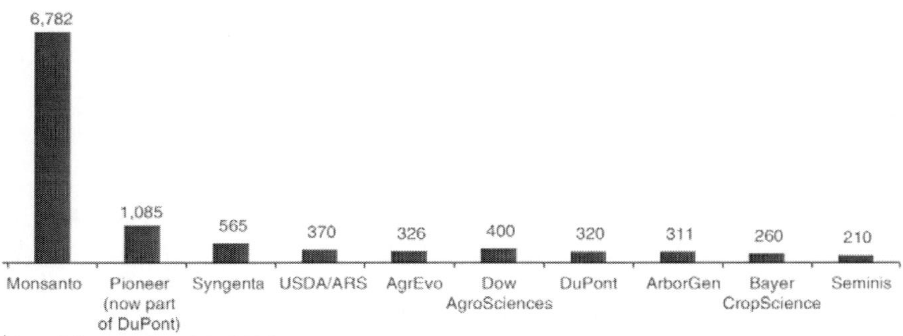

*As of September 24, 2013.

Authorizations for field releases of GE plant varieties are issued by USDA's Animal and Plant Health Inspection Service (APHIS) to allow technology providers to pursue field testing.

Source: Information Systems for Biotechnology (ISB, 2013).

Figure 4. Institutions having the most authorized permits and notifications (number held).

Table 2. Biotech crops currently available and in development

Crop	Input traits				Output traits	
	Herbicide tolerance	**Insect resistance**	**Virus/fungi, resistance**	**Agronomic properties[11]**	**Product quality[14]**	**Pharmaceuticals /nutraceuticals[17]**
Corn	C	C[5]	D	C[12] D	D	D
Soybeans	C	D		D	C[15] D	
Cotton	C	C[6]		D	D	
Potatoes		W[7]	D	D	D	D
Wheat	C[2]		D			
Other field crops[1]	C[3] D[4]	D	D	D	D	D
Tomato, squash, melon, sweet corn		C[8]	C9 D	D	C[16] D	D
Other vegetables	D				D	
Papaya			C[10]			
Fruit trees			D		D	
Other trees				D[13]	D	
Flowers					D	

[1] Includes barley, canola, peanuts, tobacco, rice, sugar beet, alfalfa, etc.

[2] Monsanto discontinued breeding and field level research on its GE Roundup Ready wheat in 2004.

[3] Canola, sugar beet, alfalfa.

[4] Barley, rice.

[5] Bt corn to control the corn borer commercially available since 1996; Bt corn for corn rootworm control commercially available since 2003; Bt corn to control the corn earworm commercially available since 2010; stacked versions of them also available.

[6] Bt cotton to control the tobacco budworm, the bollworm, and the pink bollworm, commercially available since 1996.

[7] Bt potatoes, containing built-in resistance to the Colorado potato beetle were commercially introduced in 1996 and withdrawn in 1999.

[8] Sweet corn with insect resistance (to the corn earworm and European corn borer) was planted in anout 20,000 acres and sold in the fresh market in 2008 (NRC, 2010).

[9] VR squash accounted for about 12 percent of the squash produced in in 2005 (NRC, 2010).

[10] Responding to a devastating papaya virus epidemic in the mid-1990s, researchers at Cornell University and at the University of Hawaii developed two virus-resistant varieties of GE papaya. First commercial plantings were made in 1998. The new varieties were successful in resisting a viral epidemic and were planted on more than 30 percent of Hawaii's papaya acreage in 1999.

[11] Such as resistance to drought, frost, salinity; more efficient use of nitrogen.

[12] Drought tolerant corn approved for commercial use in 2011; expected to be introduced in 2012.

[13] Modified lignin content.

[14] Includes delayed ripening (fruits and vegetables with longer shelf life); protein content, carbohydrate content, fatty acid content, micronutrient content, oil content, modified starch content, flavor and texture (fruits and vegetables), color (cotton, flowers), fiber properties (cotton), gluten content (wheat), naturally decaffeinated (coffee), and low phytase.

[15] High oleic soybeans.

[16] FlavrSavr tomato genetically engineered to remain on the vine longer and ripen to full flavor after harvest was pulled out of the market because of harvesting and marketing problems.

[17] Includes increased vitamin, iron, beta-carotene (antioxidant), lycopene (anti-cancer), amino acid content; low-calorie sugar; hypoallergenic crops; antibodies, vaccines. Industrial uses (such as specialty machine oils).

Sources: ISB (2013); Fernandez-Cornejo and Caswell (2006); National Research Council (2010); USDA Animal and Plant Health Inspection Service.

Determination of Nonregulated Status

Nonregulated status allows seed companies to commercialize the GE seeds that they have developed. After successful field testing, technology providers petition APHIS for a determination of non-regulated status. If, after review, APHIS determines that the organism (i.e., GE plant) is unlikely to pose a plant pest risk, the organism is deregulated (see box, "Regulatory Oversight") and can be moved and planted without APHIS oversight. As of September 2013, APHIS had received 145 petitions for deregulation—compared with 103 petitions received in 2005—and had granted 96 (31 were withdrawn, 17 were pending, and 1 was incomplete) (Information Systems for Biotechnology, 2013). For corn, 30 petitions were granted nonregulated status; 15 were granted for cotton; 11 for tomatoes; 12 for soybeans; 8 for canola/rapeseed; 5 for potatoes; 3 for sugarbeet; 2 each for papaya, rice, and squash; and 1 each for alfalfa, plum, rose, tobacco, flax, and chicory. By trait, as of September 2013, 43 petitions were granted nonregulated status for herbicide tolerance, 31 for insect resistance, 17 for product quality, 9 for agronomic properties, 8 for virus resistance, and 2 for others.[7]

The Research and Development Pipeline

APHIS approval for field testing and determination of nonregulated status signals that the GE products are near commercial status. In addition to crops with improved pest management traits, APHIS approvals include crops with traits that provide viral/fungal resistance; favorable agronomic properties (resistance to cold, drought, frost, salinity, more efficient use of nitrogen, increased yield); enhanced product quality such as delayed ripening, flavor, and texture (fruits and vegetables); increased protein or carbohydrate content, fatty acid content or micronutrient content; modified starch, color (cotton, flowers), fiber properties (cotton) or gluten content (wheat); naturally decaffeinated (coffee); nutraceuticals (added vitamins, iron, antioxidants such as beta-carotene); and pharmaceuticals (table 2).[8] Additional information is found in the Pew Initiative (2001), Runge and Ryan (2004), Monsanto (2012), and Pioneer (2012).

Adoption of GE Crops by U.S. Farmers

When farmers adopt a new technology, they typically expect benefits like increased farm net returns, time savings (by making farming less effort intensive), or reduced exposure to chemicals. Net benefits are a function of farm characteristics and location, output and input prices, existing production systems, and farmer abilities and preferences.

Judging by the widespread adoption of GE seeds, farmers have benefited from them. U.S. farmers planted about 169 million acres of GE corn, soybeans, and cotton in 2013 (table 3), accounting for almost half of the estimated total land used to grow all U.S. crops.

On a global scale, approximately 420 million acres of GE crops were planted in 28 countries in 2012 (International Service for the Acquisition of Agri-biotech Applications, 2012). U.S. acreage accounted for approximately 41 percent of acres planted with GE seed, Brazil accounted for 21 percent, Argentina for 14 percent, Canada for 7 percent, India for 6 percent, and China, Paraguay, South Africa, and Pakistan each for roughly 2 percent.

Table 3. Major genetically engineered crops, 2000-2013

Year	GE corn		GE soybeans		GE cotton	
	Million acres planted	Percent of corn acres	Million acres planted	Percent of soybean acre	Million acres planted	Percent of cotton acres
2000	19.89	25	40.10	54	9.47	61
2001	19.68	26	50.37	68	10.88	69
2002	26.82	34	55.47	75	9.91	71
2003	31.44	40	59.46	81	9.84	73
2004	38.04	47	63.93	85	10.38	76
2005	42.53	52	62.67	87	11.25	79
2006	47.78	61	67.21	89	12.68	83
2007	68.27	73	58.91	91	9.42	87
2008	68.79	80	69.66	92	8.15	86
2009	73.42	85	70.48	91	8.05	88
2010	75.85	86	71.99	93	10.21	93
2011	81.21	88	70.46	94	13.25	90
2012	85.50	88	71.79	93	11.58	94
2013	87.64	90	72.29	93	9.23	90

Genetically engineered crops in this table include Bt crops that have insect-resistant traits or HT crops that have herbicide tolerance traits, or both.

Sources: USDA Economic Research Service using data from from USDA/NASS Quick Stats and Fernandez-Cornejo (2013).

Commercially introduced in the United States in 1996, major GE crops were rapidly adopted. Planting of GE crops (measured in acres) increased by 68 percent between 2000 and 2005 and grew by 45 percent between 2005 and 2013. Three crops (corn, cotton, and soybeans) make up the bulk of U.S. acres planted to GE crops (table 3), mostly for herbicide tolerance (HT) and insect resistance (Bt). Including varieties with HT and/or Bt traits, GE crops accounted for 90 percent of all planted cotton acres, 93 percent of soybean acres, and 90 percent of corn acres in 2013. U.S. farmers have tended to adopt HT seeds at higher levels than seeds with insect resistance (figure 5). In part, this is because weeds are a pervasive problem.[9] HT adoption was particularly rapid in soybeans, with U.S. farmers increasing their planting of HT soybeans from 54 percent of soybean acres in 2000 to 87 percent in 2005 and 93 percent in 2013. HT cotton increased from 46 percent of cotton acres in 2000 to 61 percent in 2005 and 82 percent in 2013. HT corn increased from 7 percent of corn acres in 2000 to 26 percent in 2005 and 85 percent in 2013. Insect infestations tend to be more localized than weed infestations (figure 6). Farmers planted Bt cotton (to control insects such as tobacco budworm, cotton bollworm, and pink bollworm) on 35 percent of the cotton acres in 2000, 52 percent in 2005, and 75 percent in 2013. Bt corn—commercially introduced to control the European corn borer in 1996, the corn rootworm in 2003, and the corn earworm in 2010—was planted on 19 percent of corn acres in 2000, 35 percent in 2005, and 76 percent in 2013.

Other GE crops commercially grown in the United States are HT canola, HT sugarbeets, HT alfalfa, virus-resistant papaya, and virus-resistant squash.[10] In addition, other traits are being developed and tested, including cold/drought resistance and enhanced protein, oil, or vitamin content (see table 2).[11]

Based on the Agricultural Resource Management Survey (see box, "ARMS Data"),[12] farmers indicate that they adopted GE corn, cotton, and soybeans primarily to increase yields (figure 7). Other popular reasons for adopting GE crops were to save management time, to facilitate other production practices (such as crop rotation and conservation tillage), and to reduce pesticide costs.

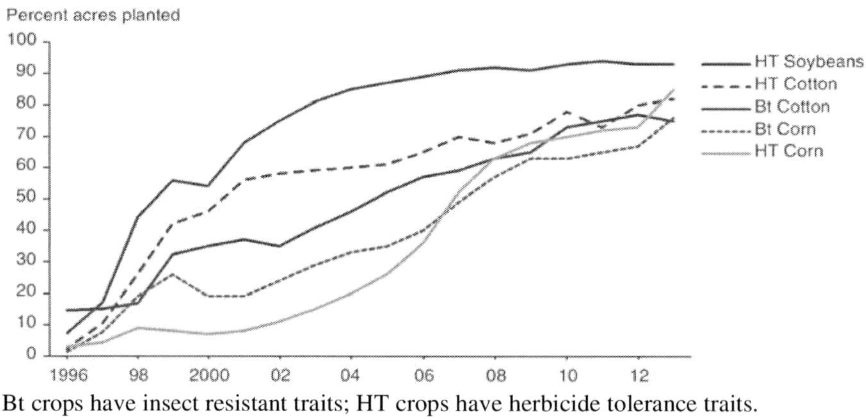

Bt crops have insect resistant traits; HT crops have herbicide tolerance traits.
Data for each crop category include varieties with both Bt and HT (stacked) traits.
Source: U.S. Department of Agriculture (USDA), Economic Research Service (ERS). 2013. *Adoption of Genetically Engineered Crops in the U.S.* data product.

Figure 5. Adoption of genetically engineered crops in the United States.

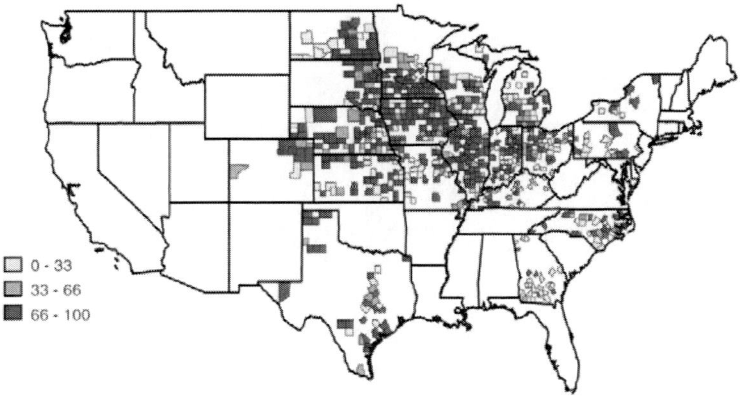

Bt crops have insect-resistant traits.
Source: USDA, Economic Research Service using data from the 2010 Agricultural Resource Management Survey (ARMS) Phase II corn survey.

Figure 6. Percentage of U.S. corn farmers who adopted Bt seeds in 2010.

THE ARMS DATA

The Agricultural Resource Management Survey (ARMS), sponsored by USDA's National Agricultural Statistics Service (NASS) and the Economic Research Service (ERS), has a multiphase, multi-frame, stratified, probability-weighted design. In other words, farmers with preselected characteristics are administered the ARMS each year. After data collection, NASS generates probability weights to help ensure that the ARMS sample accurately represents the population of U.S. famers.

The ARMS has three phases. Phase I, administered in the summer of the survey year, verifies that all respondents operate a farm or plant a specific crop. Phase II, administered in the fall or winter of the survey year, is a field-level survey that collects data on production practices and input use. Phase III, administered in the spring following the survey year, gathers data on debt, revenue, operating costs, and expenditures for the entire farm.

Phase I and Phase III are annual surveys that are administered to all respondents. Phase II is a commodity-specific survey that is administered annually for a rotating selection of crops. For instance, the ARMS Phase II Corn survey was administered in 2005 and 2010.

The Phase II Soybean survey was administered in 2006 and the cotton survey was administered in 2007.

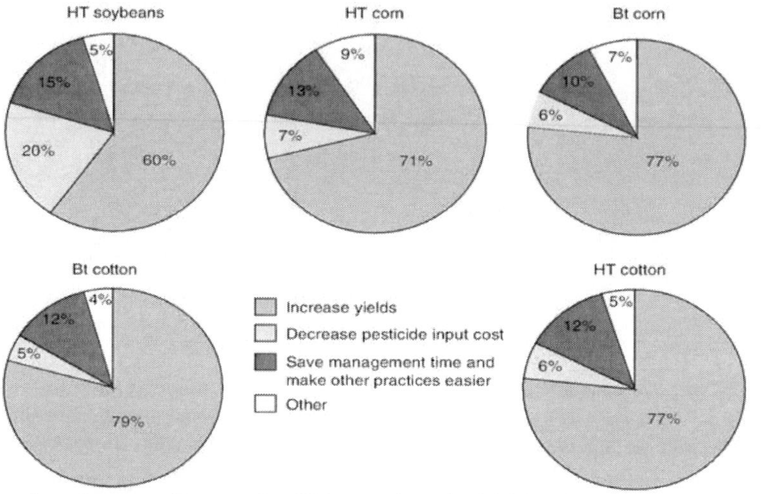

Bt crops have insect resistant traits; HT crops have herbicide tolerance traits.

Sources: USDA Economic Research Service using data from Agricultural Resource Management Survey (ARMS) Phase II surveys: 2010 for corn, 2007 for cotton, and 2006 for soybeans.

Figure 7. Farmers' reasons for adopting genetically engineered crops.

Farm-Level Economic Impacts of GE Crop Adoption

The impacts of GE crop adoption vary by crop and technology. Many studies have assessed the factors that influence adoption as well as the impacts of GE crops on yields, net returns, and pesticide use (table 4; Fernandez-Cornejo and McBride, 2002). Over the first 15 years of commercial use, GE seeds have not been shown to increase yield potentials of the varieties.[13] In fact, the yields of herbicide-tolerant or insect-resistant seeds may be occasionally lower than the yields of conventional varieties if the varieties used to carry the HT or Bt genes are not the highest yielding cultivars, as in the earlier years of adoption (Fernandez-Cornejo and Caswell, 2006; National Research Council, 2010).[14] However, by protecting the plant from certain pests, GE crops can prevent yield losses to pests, allowing the plant to approach its yield potential.

Table 4. Summary of selected studies on the effects of genetically engineered crops on yields, pesticide use, and net returns

Crop/researchers/date of publication	Effects on			
	Data source	Yield	Pesticide use	Net returns
Herbicide-tolerant soybeans				
Delannay et al., 1995	Experiments	Same	na	na
Roberts et al., 1998	Experiments	Increase	Decrease	Increase
Arnold et al., 1998	Experiments	Increase	na	Increase
Marra et al., 1998	Survey	Increase	Decrease	Increase
Reddy and Whiting, 2000	Experiments	Same	na	Increase
Duffy, 2001	Survey	Small decrease	na	Same
Fernandez-Cornejo et al., 2002[1]	Survey	Small increase	Small increase	Same
McBride & El-Osta, 2002[2]	Survey	na	na	Same
Bradley et al., 2004	Experiments	Same	na	na
Marra et al., 2004	Survey	Same	na	Increase
Herbicide-tolerant cotton				
Vencill, 1996	Experiments	Same	na	na
Keeling et al., 1996	Experiments	Same	na	na
Goldman et al., 1998	Experiments	Same	na	na
Culpepper and York, 1998	Experiments	Same	Decrease	Same
Fernandez-Cornejo et al., 2000[1]	Survey	Increase	Same	Increase
Adhicari et al. 2000	Survey	na	na	Increase
Herbicide-tolerant corn				
Fernandez-Cornejo and Klotz-Ingram, 1998	Survey	Increase	Decrease	Same
Ferrell and Witt, 2002	Experiments	Same	na	Small increase
McBride & El-Osta, 2002[2]	Survey	na	na	Increase
Parker et al., 2006	Experiments	Same	na	na

Table 4. (Continued)

Crop/researchers/date of publication	Data source	Effects on Yield	Effects on Pesticide use	Effects on Net returns
Bt cotton				
Stark, 1997	Survey	Increase	Decrease	Increase
Gibson et al., 1997	Survey	Increase	na	Increase
ReJesus et al., 1997	Experiments	Same	na	Increase
Bryant et al., 19993	Experiments	Increase	na	Increase
Marra et al., 1998	Survey	Increase	Decrease	Increase
Fernandez-Cornejo et al., 2000[1]	Survey	Increase	Decrease	Increase
Falck-Zepeda et al., 2000[1]	Survey	Increase	na	Increase
Cattaneo et al., 2006	Survey	Increase	Decrease	na
Piggott and Marra, 2007	Experiments	Increase	na	Increase
Bt corn				
Rice and Pilcher, 1998[1]	Survey	Increase	Decrease	Depends on infestation
Marra et al., 1998	Survey	Increase	Decrease	Increase
Duffy, 2001[2]	Survey	Increase	Na	Same
Baute, Sears, and Schaafsma, 2002	Experiments	Increase	Na	Depends on infestation
McBride & El-Osta, 2002[4]	Survey	Na	Na	Decrease
Pilcher et al., 2002[5]	Survey	Increase	Decrease	Na
Dillehay et al., 2004[6]	Experiments	Increase	Na	Na
Mitchell et al., 2004[7]	Experiments	Increase	Na	Depends on infestation
Fernandez-Cornejo and Li, 2005[8]	Survey	Increase	Decrease	Na
Mungai et al., 2005[9]	Experiments	Increase	Na	Na
Fang et al., 2007[10]	Experiments	Increase	Na	Na

na = not analyzed in the study; [1]Results using 1997 data; [2]Results using 1998 data; [3]Results are for 1996 and 1998, results were different for 1997 when the pest pressure was low; [4]Results using 1998 data; [5]Results using 1996-1998 data; [6]Results using 2004-2006 data; [7]Results using data from 1997-1999 ; [8]Results using data from 2001; [9]Results using data from 2002-2003, [10]Results using data from 2002.

The profitability of GE seeds for individual farmers depends largely on the value of the yield losses mitigated and the associated pesticide and seed costs.[15] GE adoption tends to increase net returns if the value of yield losses mitigated plus the pesticide savings exceeds the additional GE seed costs.

Adoption of Bt crops increases yields by mitigating yield losses to pests. Bt crops are particularly effective at mitigating yield losses. For example, before Bt corn was commercially introduced in 1996, the European corn borer was only partially controlled using chemical insecticides (Fernandez-Cornejo and Caswell, 2006). Chemical use was not always profitable, and timely application was difficult. Many farmers accepted expected yield losses of 0.4 to 3.2 bushels from this pest rather than incur the expense and uncertainty of chemical control (Hyde et al., 1999). After the introduction of Bt corn, adopters who had previously controlled corn borer infestations using insecticides lowered their pesticide costs and increased their yields. Adopters who had not previously treated European corn borer infestations with insecticides achieved only yield gains (and may have incurred higher seed costs).

In addition to improvements in background germplasm, Bt corn yields have increased over time as new insect resistance traits have been incorporated into the seeds and multiple (stacked) traits have become available (Fernandez-Cornejo and Wechsler, 2012). For instance, upon commercial introduction in 1996, Bt corn seeds were only resistant to one type of pest: the European corn borer. Since then, resistance to corn rootworms (2003) and corn earworms (2010) has been introduced.

Most experimental field tests and farm surveys show that Bt crops produce higher yields than conventional crops (table 4). Intuitively, Bt adopters are more likely to obtain higher yields than nonadopters by controlling insects and thus reducing yield losses to pests. The yield gain of Bt crops has become larger in recent years as new Bt traits have been incorporated into the seeds and multiple (stacked) traits have become available. For example, ARMS data show that the yield gain by Bt corn adopters relative to conventional varieties increased from 12.5 bushels per acre in 2001 to 16 bushels in 2005 and 26 bushels in 2010 (table 5; Fernandez-Cornejo and Li, 2005).[16] The geographical distributions of Bt adoption rates and average corn yields for 2010 are shown in figures 6 and 8, respectively.

Table 5. Bt corn adopters and non-adopters, 2005 and 2010 (Sample means of selected variables)

Variable	Unit	Bt	Non -Bt	Difference	Significance
2005					
Yield	Bushels/acre	155.1	138.6	16.6	***
Insecticide use	Pounds Ai/acre	0.05	0.09	-0.04	1
Corn price	Dollars/bushel	1.95	2.01	-0.06	NS
2010					
Yield	Bushels/acre	159.2	132.7	26.5	***
Insecticide use	Pounds Ai/acre	0.02	0.02	0.00	NS
Corn price	Dollars/bushel	5.39	5.40	-0.01	NS

*, **, and *** Indicates statistical significance at 10, 5, and 1 percent level, respectively.

NS = Not significant. [1]Significant at the 5-percent level when using standard procedures but not significant (p value 0.15) when using the delete-a-group jacknife procedure to estimate variances (Kott, 1998).

Source: USDA Economic Research Service using data from 2005 and 2010 Agricultural Resource Management Survey corn surveys.

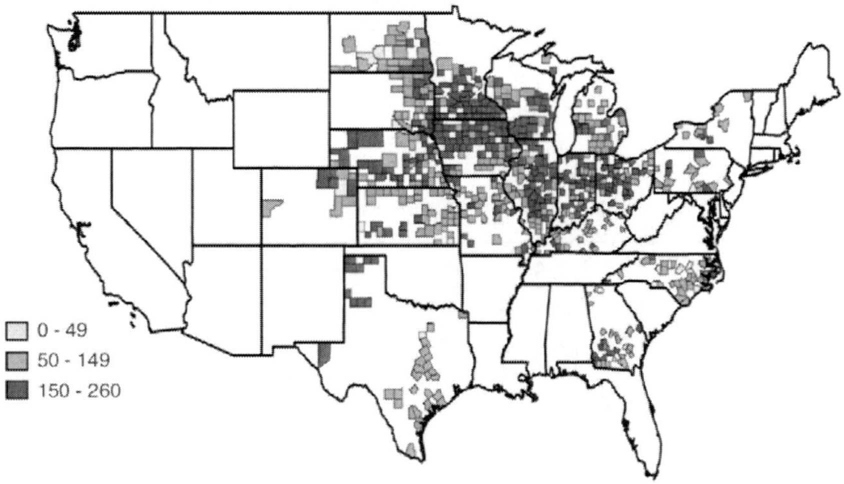

Source: USDA, Economic Research Service using data from 2010 Agricultural Resource Management Survey (ARMS) Phase II corn survey.

Figure 8. Average yields (in bushels per acre) for U.S. corn farmers in 2010.

While mean comparisons are illustrative, definitive conclusions about relative yields are possible only if the data are generated under experimental settings where factors other than adoption are controlled for by making them as similar as possible (Fernandez-Cornejo and McBride, 2002; NRC, 2010).[17] This is not the case with survey data.[18] Bt use is not random. Surveyed farmers are not randomly assigned to a treatment group (adopters) and a control group (nonadopters). Consequently, adopters and nonadopters may be systematically different from one another (for example, in terms of management ability). If these differences affect both farm performance and Bt adoption, they will confound the analysis (Fernandez-Cornejo and McBride, 2002; Fernandez-Cornejo et al., 2002). This self-selection[19] biases the statistical results unless it is corrected (Greene, 1997). Fernandez-Cornejo and Wechsler (2012) specified an econometric model to estimate the impact of adoption that accounts for self-selection. Using this model, they found that a 10-percent increase in the probability of adopting Bt corn was associated with a 1.7-percent increase in yields in 2005, and in a new ERS analysis using 2010 survey data, they found a 2.3-percent increase in yields (table 6). Using a similar econometric method to analyze cotton data, ERS researchers found that a 10-percent increase in the probability of adopting Bt cotton was associated with a 2.1-percent increase in yields in 1997 (Fernandez-Cornejo and McBride, 2002).

The effect of HT seeds on yields is mixed. The evidence on the impact of HT seeds on soybean, corn, and cotton yields is mixed (table 4). Several researchers found no significant difference between the yields of adopters and nonadopters of HT; some found that HT adopters had higher yields, while others found that adopters had lower yields. For instance, an ERS study found that a 10-percent increase in the adoption of HT cotton led to a 1.7-percent increase in cotton yields. HT soybean adoption was associated with a statistically significant, but small, increase in yields: a 10-percent increase in the probability of adopting HT soybeans was associated with a 0.3-percent increase in yields (Fernandez-Cornejo and McBride, 2002).

ARMS results show that HT soybean yields were 5 bushels per acre (3 percent) higher than conventional soybean yields in 2006 (but only significantly different at the 10-percent level) (table 7). In the case of corn, ARMS results show that HT corn yields were similar to those of conventional corn in 2010. However, unlike soybeans, the majority of corn (and cotton) producers in recent years use seed with stacked traits (figures 9 and 10). Multiple stacked traits make evaluating the effect of individual GE traits on yields and profitability more complicated.

Table 6. The Impact of adopting Bt corn: Elasticities 2005, 2010[1]

Variable	Elasticity with respect to the probability of adoption	
	2005	2010
Net returns	0.17	0.23
Yield	0.17	0.23
Seed	0.1	0.21
Insecticide	NS	NS

[1] Elasticity measures the responsiveness of a variable (e.g., s, yield) to a change in another (e.g., adoption rate). It is unit free and always expressed in percentage terms.

Bt crops have insect-resistant traits

NS = Not significant.

Sources: 2005: Fernandez-Cornejo and Wechsler (2012). 2010: New analysis by Economic Research Service. (Model results using 2010 ARMS corn data. Model specification similar to that used by Fernandez-Cornejo and Wechsler, 2012).

Table 7. HT soybean adopters and non-adopters, 2006

Variable	Units	HT adopters	Non-adopters	Difference	Significance
Yield	Per acre yields, in bushels	45.6	40.6	5.0	*
Total herbicide use	Pounds AI per acre	1.36	1.05	0.31	NS
Glyphosate use	Pounds per acre	1.23	0.38	0.85	***
Other herbicides use	Pounds per acre	0.13	0.66	-0.53	**1

*, **, and *** Indicates statistical significance at 10-, 5-, and 1-percent level, respectively. NS = Not significant.

[1] Significant at the 5-percent level when using standard procedures but not significant (p value = 0.14) when using the jacknife procedure to estimate variances (Kott, 1998). HT crops have herbicide tolerance traits.

Source: Economic Research Service using data from 2006 Agricultural Resource Management Survey soybean survey.

Stacked-trait seeds tend to have higher yields. An analysis of ARMS corn data indicates that stacked seeds (seeds with several GE traits) have higher yields than conventional seeds or seeds with only one GE trait. For example, 2010 ARMS data show that conventional corn seeds had an average yield of

134 bushels per acre in 2010. By contrast, seeds with two types of herbicide tolerance (glyphosate and glufosinate) and three types of insect resistance (corn borer, corn rootworm, and corn earworm) had an average yield of 171 bushels per acre. These results are consistent with findings by Nolan and Santos (2012), who analyzed a rich dataset of experimental hybrid trials collected by the extension services of 10 universities in major corn-producing States from 1997 to 2009.

Not surprisingly, adoption rates of stacked-seed varieties have increased quickly (figures 9 and 10). Stacked corn seeds grew from 1 percent of the corn acres in 2000 to 9 percent in 2005 and 71 percent in 2013, while stacked cotton seeds grew from 20 percent to 34 percent in 2005, and 67 percent in 2013 (figures 9-10). The most widely adopted GE corn varieties have both Bt and HT traits (table 8). Varieties with three or four traits are now common.

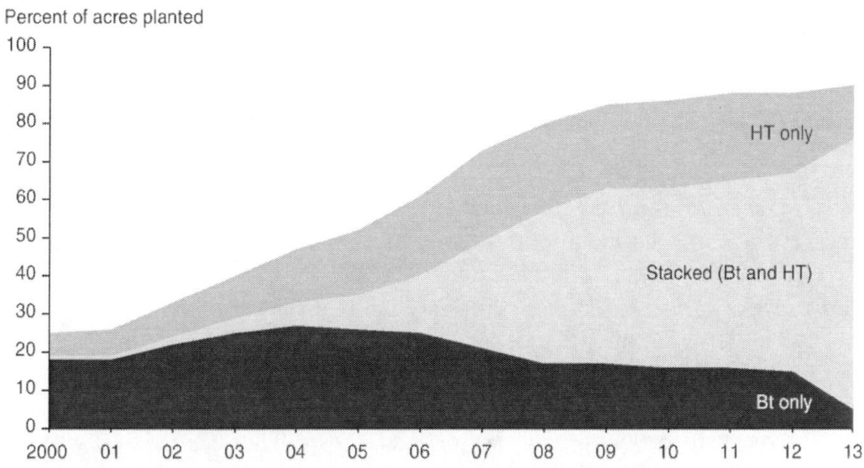

Bt crops have insect-resistant traits; HT crops have herbicide tolerance traits.

Source: U.S. Department of Agriculture (USDA), Economic Research Service (ERS). 2013. *Adoption of Genetically Engineered Crops in the United States,* data product.

Figure 9. Adoption of genetically engineered corn: growth of stacked traits, 2000-2013.

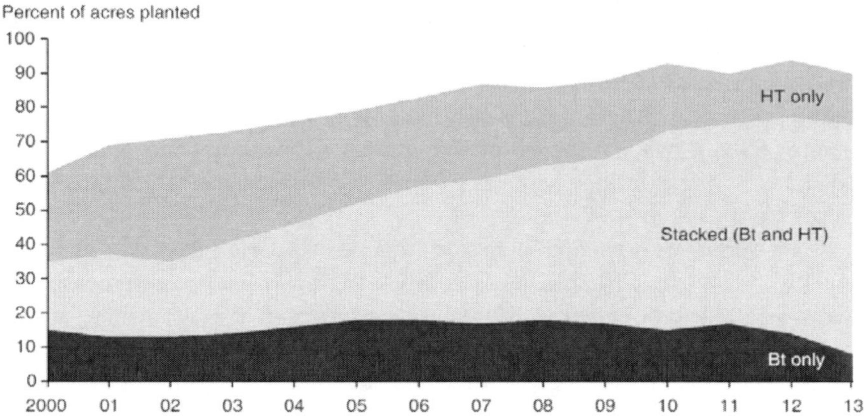

Bt crops have insect-resistant traits; HT crops have herbicide tolerance traits.

Source: U.S. Department of Agriculture (USDA), Economic Research Service (ERS). 2013. *Adoption of Genetically Engineered Crops in the United States,* data product.

Figure 10. Adoption of genetically engineered cotton: growth of stacked traits, 2000-2013.

GE seed prices are influenced by stacking and many other factors. The market price of seed incorporates the costs associated with seed development, production, marketing, and distribution (Fernandez-Cornejo, 2004). The price must reflect farmers' willingness to pay while ensuring a profit margin after costs. Furthermore, the price depends on the competitiveness of the particular seed market, and the pricing behavior of those firms that hold large shares of the market (NRC, 2010).

In recent decades, private sector R&D costs have been rising with the application of new technologies, and much of the increase in seed prices has been associated with this trend (Krull et al., 1998). R&D costs vary among the different seed markets. For example, the corn seed market depends extensively on private sector R&D and passes these costs on to farmers. The wheat seed market depends largely on public sector research, which is largely cost free for farmers. There is no GE wheat commercially available.[20]

The real price index for seed rose nearly 30 percent faster than the average index of prices paid by U.S. farmers over 1996-2007 (NRC, 2010). The price of GE soybean and corn seeds grew by about 50 percent in real terms (adjusted for inflation) between 2001 and 2010 (figure 11). The price of GE cotton seed grew even faster (NRC, 2010).

Table 8. Adoption of genetically engineered varieties by U.S. corn producers, 2010

Seed Type	GE traits (percent adopters)			
	Bt only	HT only	Bt/ HT	No GE
1. Genetically modified **herbicide resistant** seed variety (*e.g. LIBERTYLINK; ROUNDUP READY CORN*)		21.36		
2. Non-genetically modified **herbicide resistant** seed variety (*e.g. IMI-CORN*)				3.48
3. Genetically-modified **Bt** variety for insect resistance to control the European Corn Borer (**Bt-ECB**) (*e.g. YIELDGARD, YIELDGARD CORN BORER, HERCULEX I, NATUREGARD, KNOCKOUT*)	7.12			
4. Genetically modified **Bt** variety for insect resistance to control the corn rootworm (**Bt-CRW**) (*e.g. YIELDGARD ROOT-WORM, HERCULEX RW*)	3.06			
5. **Stacked** gene (trait) variety with both genetically modified **Bt-ECB** and **Bt-CRW** (*e.g. YIELDGARD PLUS, HERCULEX XTRA*)	3.81			
6. **Stacked** gene variety with **two** genetically modified herbi-cide resistant traits (*e.g. LIBERTYLINK + ROUNDUP READY*)		3.73		
7. **Stacked** gene variety with both genetically modified **Bt-ECB** and **herbicide resistant** (*e.g. YIELDGARD + ROUNDUP READY, YIELDGARD CORN BORER WITH ROUNDUP READY CORN 2, HERCULEX I + LIBERTYLINK*)			9.77	
8. **Stacked** gene variety with both genetically modified **Bt-CRW** and **herbicide resistant** (*e.g. YIELDGARD ROOT-WORM WITH ROUNDUP READY CORN 2, HERCULEX CW + ROUNDUP READY CORN*)			8.03	
9. **Triple stacked** gene variety with genetically modified **Bt-ECB** and **Bt-CRW** plus **herbicide resistant** traits (*e.g. YIELD-GARD PLUS WITH ROUNDUP READY CORN 2, HERCULEX XTRA + LIBERTYLINK*)			25.91	

Table 8. (Continued)

Seed Type	GE traits (percent adopters)			
	Bt only	**HT only**	**Bt/ HT**	**No GE**
10. **Stacked** gene varieties that, in addition to the ECB and the rootworm, can control the corn earworm			5.71	
11. Multiple (more than three) **trait** stacked variety with sev-eral Bt traits and two herbicide resistant traits—glyphosate (Roundup) and glufosinate (Liberty)			1.24	
12. None of the above				6.79
Total	13.99	25.08	50.66	10.26

Source: USDA Economic Research Service using data from 2010 Agricultural Resource Management Survey corn survey.

The increase in GE seed prices can be attributed in part to increasing price premiums over conventional seeds (which include technical fees) associated with the rising share of GE seeds with more than one trait and/or more than one mode of action for particular target pests (NRC, 2010). Another factor contributing to the increase in GE seed prices is the improvement in seed genetics (germplasm) (NRC, 2010). The rapid adoption of GE crops indicates that many farmers are willing to pay higher seed prices because of improved seed performance and the additional pest management traits embedded in the GE seed.

Various studies of stacked GE seed varieties have found that stacked seeds are priced less than the sum of their component values (Stiegert et al., 2010). Shi et al. (2008, 2010) note that sub-additive pricing is consistent with "the presence of economies of scope in seed production." Moreover, these scope economies are consistent with "synergies in R&D investment (treated as a fixed cost)" across stacked seeds that can contribute to reducing total cost (Shi et al., 2010). Shi et al. (2009) found that while increased concentration in the seed industry has contributed to higher seed prices, complementarity effects in production and distribution mitigate these effects. Kalaitzandonakes et al. (2010-11) conclude that, while estimation of market power and associated price markups is not straightforward, the U.S. seed industry show both "moderate market power" and dynamic market efficiency (as indicated by the balance between firm profits and investments in product quality and innovation) over their period of analysis (1997-2008).

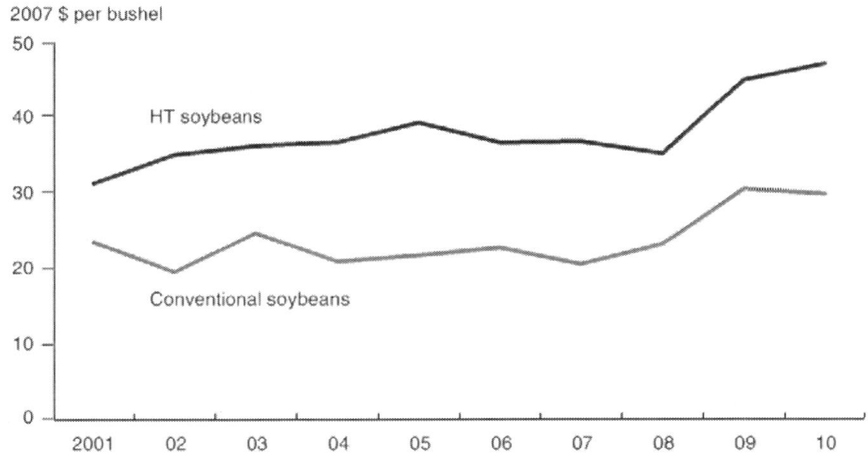

Prices of genetically engineered (GE) seed are higher than those of non-GE seed, corn

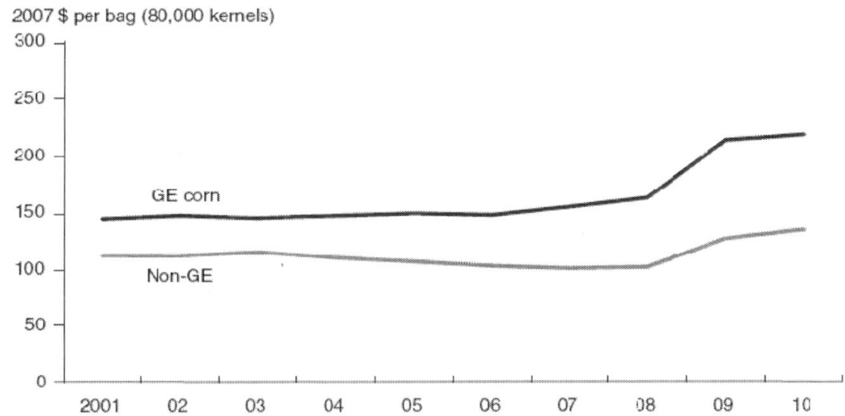

HT crops have herbicide tolerance traits.

Source: USDA Economic Research Service using data from USDA National Agricultural Statistics Service *Agricultural Prices*, various years.

Figure 11. Prices of genetically engineered (GE) seed are higher than those of non-GE seed, soybeans.

Adoption, Net Returns, and Farm Household Income

The impacts of GE crop adoption vary by crop and technology. Most studies show that adoption of Bt cotton and Bt corn is associated with increased net returns (table 4). However, some studies of Bt corn show that profitability is strongly dependent on pest infestation levels.[21] The impact of HT seeds (for corn, cotton, and soybeans) on net returns depends on many factors.

Planting Bt cotton and Bt corn is often more profitable than planting conventional seeds. ERS researchers found that adoption of Bt cotton was positively associated with net producer returns in 1997 (Fernandez-Cornejo and McBride, 2002). Using 2005 ARMS data, Fernandez-Cornejo and Wechsler (2012) found that a 10-percent increase in the probability of adopting Bt corn was associated with a 1.7-percent increase in net returns. In a new ERS analysis using 2010 ARMS data, we find that a 10-percent increase in the probability of adopting Bt corn was associated with 2.3-percent increase in net returns (table 6). Thus, there is essentially no change compared to earlier findings that planting Bt cotton and Bt corn is more profitable, as measured by net returns, than planting conventional seeds.

The effect of HT seeds on net returns depends on many factors. A primary advantage of herbicide-tolerant crops over traditional crops is cost savings (Fernandez-Cornejo and McBride, 2002). Producers who plant HT crops expect to achieve at least the same output while lowering weed control costs for chemicals and for mechanical methods, and minimizing the need for scouting. In return, producers pay more for HT seeds.

An additional economic effect is that the substitution of glyphosate, used in most herbicide-tolerant programs, for other herbicides decreases the demand for (and thus the price of) other herbicides (Fernandez-Cornejo and McBride, 2002). Thus, the introduction of HT seeds may have lowered pesticide costs for both HT seed adopters and nonadopters.

Finally, HT seed-based production programs allow growers to use one product to control a wide range of both broadleaf and grass weeds instead of using several herbicides to achieve adequate weed control. Herbicide-tolerant crops also complement ongoing trends toward post-emergence weed control, the adoption of conservation tillage practices, and the use of narrow row spacing. The simplicity and flexibility of weed control programs for HT seeds require less management attention, freeing valuable management time for other activities (Fernandez-Cornejo and McBride, 2002).

HT seed has a mixed effect on net returns. The evidence on the impact of HT seeds (for corn, cotton, and soybeans) on net returns is mixed (table 4). Several researchers (Fernandez-Cornejo and McBride, 2002; Bernard et al., 2004; Marra et al., 2002) found that the adoption of herbicidetolerant cotton has a positive impact on net returns. For example, Fernandez-Cornejo and

McBride (2002) found that the elasticity of net returns with respect to the probability of adoption of herbicidetolerant cotton was +0.18.[22] Bernard et al. (2004) found that adopting HT soybeans improved profits on Delaware farms. However, Fernandez-Cornejo et al. (2002) and McBride and El-Osta (2002) found no significant difference between the net returns of adopters and nonadopters of HT soybeans. Bullock and Nitsi (2001) found that HT soybean farmers are less profitable than their conventional counterparts. Overall, the empirical evidence on the impact of adopting herbicide-tolerant soybeans on net returns is inconclusive (NRC, 2010).[23]

The fact that several researchers found no significant differences between the net returns of adopters and nonadopters of HT crops (particularly HT soybeans) despite the rapid adoption of these crops suggests that many adopters may derive nonmonetary benefits from HT adoption. In particular, weed control for HT soybeans may be simpler, freeing up management time for leisure, enterprise growth, or off-farm income-generating activities.

HT crop adoption increases farm household income and has non-pecuniary benefits. ERS research shows that HT adoption is associated with higher off-farm household income for U.S. soybean farmers, most likely because time savings are used to generate income via off-farm employment (Fernandez-Cornejo et al., 2005). ERS researchers found that a 10-percent increase in the probability of adopting HT soybeans is associated with a 16-percent increase in off-farm household income. Household income from onfarm sources is not significantly associated with adoption of HT technology (Fernandez-Cornejo et al, 2007). These findings corroborate the notion that technology adoption is influenced by (or influences) the tradeoff between household/operator time spent in onfarm and off-farm activities. More recently, Gardner et al. (2009) confirm that genetically engineered crops lead to household labor savings in U.S. crop (corn and cotton) production. Using corn and soybean data, Marra and Piggott (2006) demonstrate that there are non-pecuniary benefits to GE crop adoption and show that farmers adopting GE crops place a monetary value on the convenience, flexibility, and increased worker safety associated with growing HT crops.

Adoption and Pesticide Use

Many studies based on field tests and farm surveys have examined the extent to which GE crop adoption affects pesticide (insecticide and herbicide) use, and most results show a reduction in pesticide use (table 4).

A National Research Council study (2010) concurred that GE crops lead to reduced pesticide use and /or lower toxicity compared to conventional crops.

Insecticide use decreases with the adoption of Bt crops. Generally, Bt adoption is associated with lower insecticide use (table 4). Pounds of insecticide (per planted acre) applied to corn and cotton crops have declined over the course of the last 15 years (figure 12). (Results for cotton in 1999-2001 were distorted because of the high application rates of the insecticide Malathion during the boll weevil eradication program.)

Insecticide use on corn farms declined most years and had an overall drop from 0.21 pound per corn planted acre of corn in 1995 (the year before Bt corn was commercially introduced) to 0.06 in 2005 and 0.02 pound in 2010 (figure 12). Insecticide use has declined for *both* Bt adopters and nonadopters in recent years. According to ARMS data, only 9 percent of all U.S. corn farmers applied insecticides in 2010.

Econometric studies by ERS researchers have also found that, except for recent years, Bt crop adoption led to decreases in insecticide use, controlling for other factors.

For example, FernandezCornejo et al. (2003) show that the adoption of Bt cotton in the Southeast region (which had higher rates of Bt adoption) was associated with lower insecticide use on cotton in 1997. After controlling for other factors, a 10-percent increase in Bt corn adoption was associated with a decrease in insecticide use of 4.1 percent in 2001 (Fernandez-Cornejo and Li, 2005).

However, Bt corn adoption was not significantly related to insecticide use in more recent years using 2005 data (Fernandez-Cornejo and Wechsler, 2012), as well as in a new ERS analysis using 2010 survey data (table 6).

Taken together, these results suggest that insect infestation levels on corn were lower in recent years than in earlier years and are consistent with findings by Hutchinson et al. (2010) that European corn borer populations have steadily declined over the last decade.

Moreover, several researchers have shown that areawide suppression of certain insects such as the European corn borer and the pink bollworm are associated with the use of Bt corn and Bt cotton, respectively (see box, "Bt Crop Adoption and Areawide Pest Suppression").

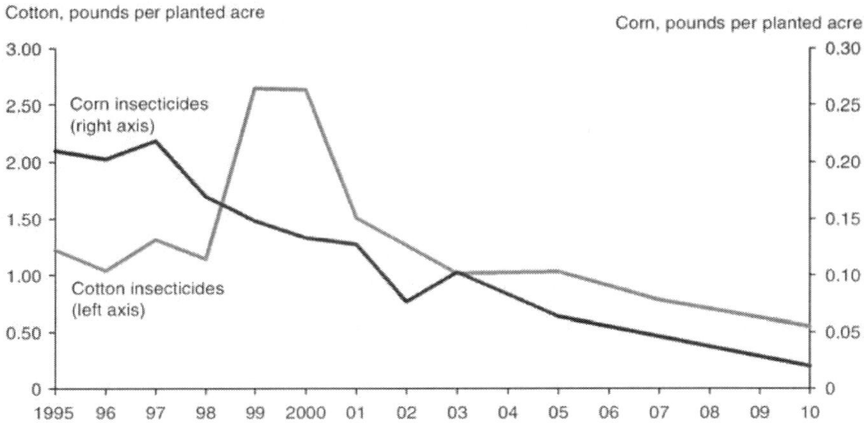

Source: USDA Economic Research Service using data from USDA National Agricultural Statistics Service Agricultural Chemical Usage reports.

Figure 12. Insecticide use in corn and cotton production, 1995-2010.

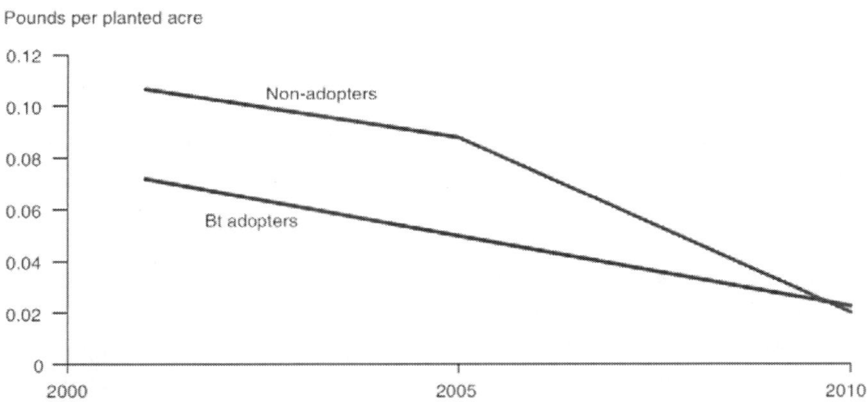

Bt crops have insect-resistant traits.

Source: USDA Economic Research Service using data from 2001, 2005, and 2010 ARMS Phase II corn surveys.

Figure 13 .Insecticide use in corn farms: adopters and non-adopters of Bt corn, 2001-2010.

Adoption of HT crops has mixed impact on herbicide use. Herbicide use on cotton and soybean acres (measured in pounds per planted acre) declined slightly in the first years following introduction of HT seeds in 1996, but increased modestly in later years (figure 14a). Herbicide use on soybean farms has been mostly constant since 1996, but increased slightly starting in 2002 and peaked in 2006. Herbicide use on corn fell from about 2.6 pounds per acre in the early years of HT corn adoption to less than 2 pounds per acre in 2002 but increased moderately in recent years. Herbicide use on corn by HT adopters increased from around 1.5 pounds per planted acre in both 2001 and 2005 to more than 2.0 pounds per planted acre in 2010, whereas herbicide use by nonadopters did not change much (figure 14b). HT adoption likely reduced herbicide use initially, but herbicide resistance among weed populations may have induced farmers to raise application rates in recent years, thus offsetting some of the economic and environmental advantages of HT corn adoption regarding herbicide use.[24]

Bt Crop Adoption and Areawide Pest Suppression

Hutchinson et al. (2010) show that areawide suppression of the European corn borer is associated with Bt corn use. They estimate that the cumulative benefits of Bt adoption over 14 years exceed $6 billion for corn growers in Illinois, Minnesota, Wisconsin, Iowa, and Nebraska. Non-adopters captured $4.3 billion of these benefits because they reap the rewards associated with low infestation rates without paying a premium for insect-resistant seeds.

Carrière et al. (2003) conducted a 10-year study in 15 regions across Arizona and showed that Bt cotton suppressed a major pest, the pink bollworm, "independent of demographic effects of weather and variation among regions." Pink bollworm population density declined only in regions where Bt cotton was abundant. Such long-term suppression has not been observed with insecticide sprays, suggesting that deployment of Bt crops may also contribute to reducing the need for insecticide sprays.

Earlier, Marra et al. (2002a) considered the side-by-side trials of Bt and conventional varieties. They discuss the bias caused by the "halo effect" that arises from the insect suppression of the Bt crops spilling over onto the conventional treatments, thus increasing the yield of the conventional crop relative to what it would be if the conventional crop were grown in isolation. This effect biases downward the yield difference between the Bt and conventional varieties.

Based on data from 36 sites in 6 provinces of northern China over 1990-2010, Lu et al. (2012) found that there was an increase in beneficial insects (such as ladybirds and lacewings) and a decrease in pests (aphids) associated with the widespread use of Bt cotton reducing insecticide sprays.

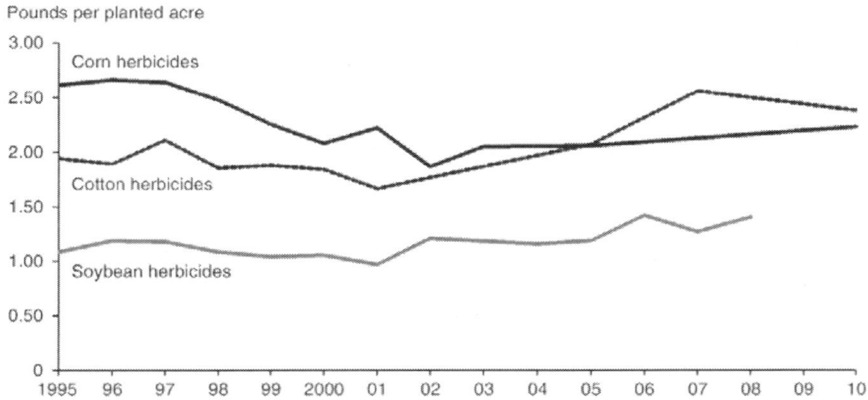

Data for herbicide use for soybeans in 2007 and 2008 are from proprietary data.

Source: USDA/NASS Agricultural Chemical Usage reports and USDA/NASS Quickstats.

Figure 14a. Herbicide use in cotton, corn, and soybeans, 1995-2010.

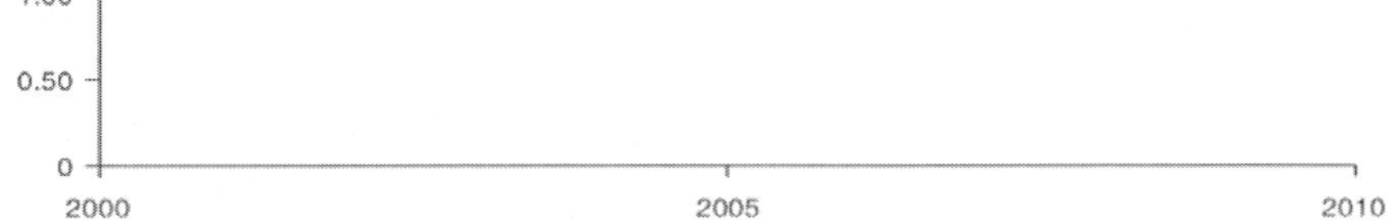

HT crops have herbicide tolerance traits.

Source: USDA Economic Research Service using data from 2001, 2005, and 2010 ARMS Phase II corn surveys.

Figure 14b. Herbicide use on corn: HT adopters and nonadopters, 2001-2010.

The main effect of HT crop adoption on herbicide use is the substitution of glyphosate for more toxic herbicides. Despite the mixed but relatively minor effect HT crop adoption has had on overall herbicide usage, most researchers agree (NRC, 2010) that the main effect of HT crop adoption is the substitution of glyphosate for more traditional herbicides. Because glyphosate is significantly less toxic and less persistent than traditional herbicides (WHO, 1994; NRC, 2010),[25] the net impact of HT crop adoption is an improvement in environmental quality and a reduction in the health risks associated with herbicide use (even if there are slight increases in the total pounds of herbicide applied).[26] However, glyphosate resistance among weed populations in recent years may have induced farmers to raise application rates. Thus, weed resistance may be offsetting some of the economic and environmental advantages of HT crop adoption regarding herbicide use. Moreover, herbicide toxicity may soon be negatively affected (compared to glyphosate) by the introduction (estimated for 2014) of crops tolerant to the herbicides dicamba and 2,4-D.

Adoption and Conservation Tillage

Conservation tillage (including no-till, ridge-till, and mulch-till) is known to provide environmental benefits (USDA's ERS/NRCS, 1998; NRC, 2010). By leaving substantial amounts of crop residue (at least 30 percent) covering the soil surface after planting, conservation tillage reduces soil erosion by wind and water, increases water retention, and reduces soil degradation and water/chemical runoff. In addition, conservation tillage reduces the carbon footprint of agriculture.

Adopters of HT crops practice conservation tillage more than growers of conventional varieties. Since the 1980s, the adoption of conservation tillage practices by U.S. farmers has been facilitated by the availability of post-emergent herbicides that can be applied over a crop during the growing season. Post-emergent herbicides are especially beneficial in no-till production systems because these herbicides control weeds without tilling the soil. HT crops have helped spread no-till farming further since they often allow a more effective system than just using other post-emergent herbicides (Fernandez-Cornejo and Caswell, 2006).

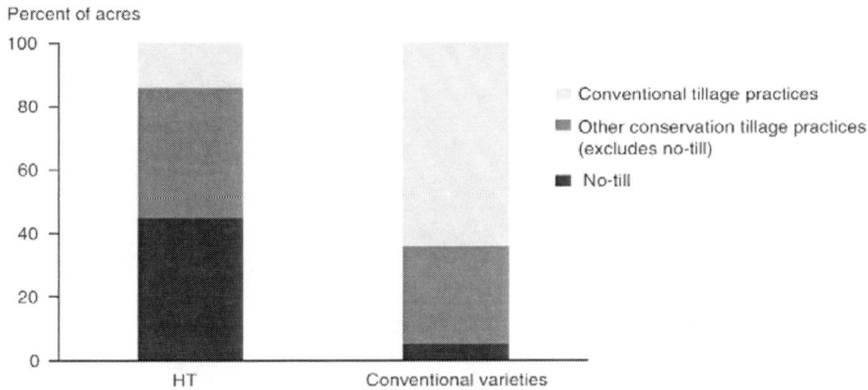

Conservation tillage includes no-till, ridge-till and mulch-till.

Source: USDA Economic Research Service using data from 2006 ARMS Phase II soybean survey.

Figure 15. Adopters of herbicide-tolerant crops used conservation tillage more than did growers of conventional varieties: soybeans, 2006.

According to USDA survey data, 60 percent of HT soybean planted acres used conservation tillage practices in 1997 versus 40 percent of conventional soybean acres (Fernandez-Cornejo and Caswell, 2006). By 2006, approximately 86 percent of HT soybean planted acres were under conservation tillage compared to only 36 percent of conventional soybean acres (figure 15).

Differences in the use of no-till specifically are just as pronounced. While approximately 45 percent of HT soybean acres were cultivated using no-till technologies in 2006, only 5 percent of the acres planted with conventional seeds were cultivated using no-till techniques.[27] Cotton and corn data exhibit similar though less pronounced patterns. Thirty-two percent of HT cotton acres were planted using conservation tillage in 2007, compared to 17 percent of conventional cotton acres (figure 16). Thirty-three percent of HT corn acres were planted using no-till in 2005, versus 19 percent of conventional corn acres (figure 17).

These trends suggest that HT crop adoption may encourage soil conservation practices. In addition, a review of several econometric studies point to a two-way causal relationship between the adoption of HT crops and conservation tillage (NRC, 2010). This implies that the adoption of herbicide-tolerant crops indirectly benefits the environment.

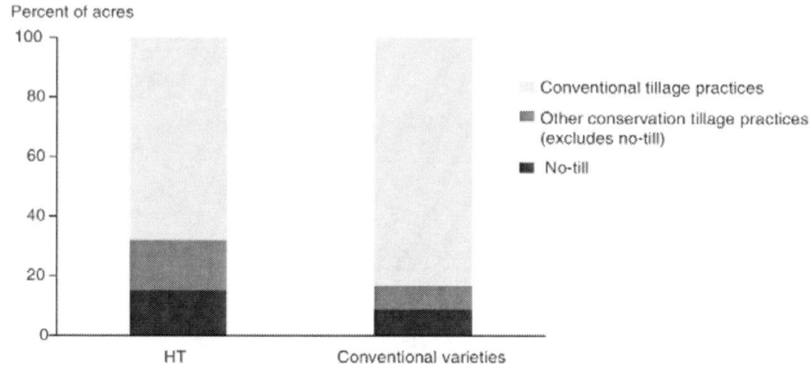

Conservation tillage includes no-till, ridge-till and mulch-till.

Source: USDA Economic Research Service using data from 2007 ARMS Phase II cotton survey.

Figure 16. Adopters of herbicide-tolerant crops used conservation tillage more than did growers of conventional varieties: cotton, 2007.

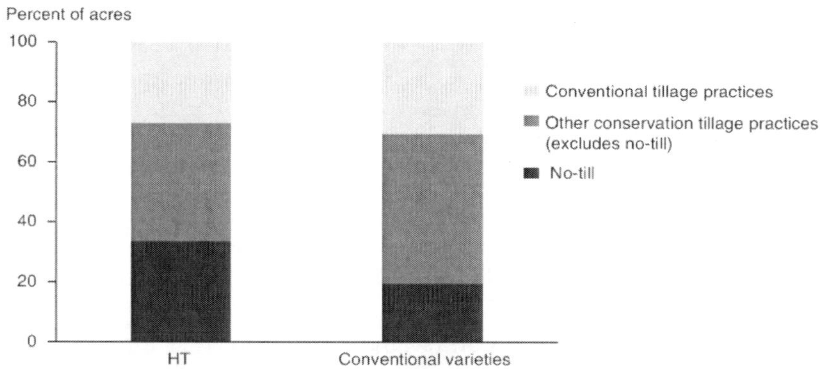

Conservation tillage includes no-till, ridge-till and mulch-till.

Source: USDA Economic Research Service using data from 2005 ARMS Phase II corn survey.

Figure 17. Adopters of herbicide-tolerant crops used conservation tillage more than did growers of conventional varieties: corn, 2005.

PEST RESISTANCE MANAGEMENT AND GE CROPS

Pesticide resistance evolution occurs when pesticide use favors the survival of pests naturally resistant to the pesticide. Over time, these resistant pests become predominant in the pest population (see box, "Economics of Resistance Management"). Developers of Bt crops and other researchers recognized early on that insect resistance to Bt toxins could develop. Measures to delay the onset of such resistance (such as refuges) were taken and, so far, the emergence of insect resistance to Bt crops has been low and of "little economic and agronomic significance" (NRC, 2010), but there are some indications that insect resistance is developing to some Bt traits in some areas.[28] Also, since many users of HT crops rely solely on glyphosate to control weeds, resistance to this herbicide was anticipated by some researchers. Thus far, overreliance on glyphosate and a reduction in the diversity of weed management practices adopted by crop producers have contributed to the evolution of glyphosate resistance in 14 weed species and biotypes in the United States.

Insect Resistance to Bt Crops

Prior to the availability of Bt crops, entomologists and other scientists successfully argued that mandatory refuge requirements—planting sufficient acres of the non-Bt crop near the Bt crop—were needed to reduce the rate at which targeted insect pests evolved resistance. Such refuges slow the rate at which Bt resistance evolves by allowing target insects that are susceptible to the Bt toxin to survive and reproduce. To be effective, the refuge must be positioned appropriately and be large enough to ensure that insects that survive on the Bt acres mate with insects that survive on the non-Bt acres. Such interbreeding increases the chances that their progeny are susceptible, having inherited Bt resistance as a recessive trait.[29] The U.S. Environmental Protection Agency (EPA) instituted mandatory refuge requirements as a condition of the registration of Bt corn and Bt cotton varieties for commercial use in the United States. This was the first time regulations were used to manage resistance to a pest control technology. Bt crop growers were required to sign a contract with their technology provider to comply with minimum refuge requirements, and technology providers were required to monitor and enforce grower compliance. An analysis of more than a decade of monitoring data suggests that the minimum refuge requirement, as well as natural refuges that also serve as hosts for target insect pests, has helped delay the evolution of Bt resistance (Tabashnik et al., 2008).

Economics of Resistance Management

When a pest population is confined to an individual farming operation, many of the benefits and costs associated with the farmer's pest control decisions accrue to and are borne by the farmer. In this hypothetical scenario, economic theory suggests that the pest population will be maintained at an economically efficient or socially optimal level.[1] However, when the pest moves from farm to farm, the pest control decisions made by any given farmer will affect the net returns accruing to that farmer, as well as those accruing to nearby farmers, although to a lesser extent. Because the effects of any farmer's control decisions on the regional pest population are practically negligible and because the benefits and costs associated with those effects are not borne by any given farmer (are not fully internalized), those effects might not be accounted for in the farmer's control decision. Because regional pest population dynamics are determined collectively by the decisions made by each farmer in the region, however, economic theory suggests that the pest population will not be maintained at a socially optimal level (Feder and Regev 1975).

This situation is referred to in the economics literature as a stock externality, an economic environment in which an individual ignores the impact of a decision that affects the level of a resource that is used by others (Gordon, 1954). In the presence of a stock externality, the resource might not be managed in a socially optimal manner. When the resource is a mobile pest population, Feder and Regev (1975) show how the introduction of a marginal user cost on pesticides, via a tax or a subsidy depending on the characteristics of the problem, can improve social welfare by ensuring that all of the net returns to pesticide use accrue to each user. The marginal user cost for a pesticide is the marginal expected present value of economic and environmental costs associated with the use of the pesticide, including impacts on regional pest population dynamics, impacts on the regional population dynamics of beneficial organisms that prey on the pest, and the evolution of resistance in the regional pest population to the pesticide, as well as the health effects associated with the accumulation of toxic pesticide residues and water/air pollution.

Pesticide resistance evolution is a process of artificial selection in which pesticide use favors the survival of particular insects and weeds and other pests resistant to the pesticide so that the frequency of resistant individuals in the population increases over time.

In the presence of a mobile pest, farmers might not account for the effects of their pesticide use decisions on the evolution of resistance nor for the effects on regional pest population dynamics (Miranowski and Carlson, 1986). Hueth and Regev (1974) suggest that the institution of a tax equal to the marginal user cost could improve social welfare by ensuring that the costs associated with resistance are incorporated by farmers. Regev et al. (1983) examine such a tax in another theoretical analysis; however, noting the difficulty of applying the tax in practice, they suggest pesticide-use restrictions as an alternative.

[1] For the purposes of this discussion, we are ignoring the costs associated with the accumulation of toxic pesticide residues, leaching of pesticides into surface and groundwater resources, and pesticide drift.

Refuge requirements depend on economic factors. Hurley et al. (2001) and Livingston et al. (2004) examine the characteristics of economically efficient refuge requirements for U.S. corn and cotton producers, respectively, for the single-toxin Bt corn and Bt cotton varieties. Both studies demonstrate that economic returns might be improved over the long run if corn and cotton producers comply with refuge requirements because of forestalling the onset of Bt resistance. The size of the economically efficient refuge requirement, however, was shown to depend on the length of the time horizon, the discount rate, and resistance evolution to conventional insecticides used to control target insect pests in the refuge acres. The refuge's ideal size was also shown to be extremely sensitive to how dominant the inherited Bt resistance trait is. Larger refuges are required to maintain susceptibility to Bt in target pest populations for longer time periods and when Bt resistance is inherited as a more dominant genetic trait by the target insect species. Livingston et al. (2007) provide empirical support for the relaxation of mandatory refuge requirements for farmers who plant cotton varieties that express multiple Bt toxins in areas that have sufficient sources of unstructured refuge.[30] These varieties control the target pest species much more effectively than single-toxin varieties. Also, most U.S. cotton is grown in areas with sufficient sources of unstructured refuge—including both cultivated and uncultivated crops and plants that serve as alternative hosts for the target insect pest species, particularly the cotton bollworm and the tobacco budworm—effectively eliminating the need for a structured (or minimum) refuge requirement. Cotton growers in Arizona, California, New Mexico, and west Texas are still required to plant minimum, structured refuges.

Refuge requirements are reduced for multiple-toxin Bt cotton varieties in some areas. EPA has eliminated the minimum refuge requirement for certain Bt cotton varieties that express multiple toxins in areas that appear to have sufficient unstructured refuge, but not for Bt corn varieties that express multiple toxins. The latter are less toxic to an important target pest known as the western corn rootworm, which might inherit Bt resistance as a partially dominant trait. Recently, western corn rootworm larvae were collected from Iowa Bt cornfields that showed evidence of root damage, and laboratory assays later confirmed that their progeny were less susceptible to Bt toxins (Gassmann et al., 2011). This has raised concerns about regulatory compliance and a continued need for minimum refuge requirements for Bt corn growers.

Evolution of Glyphosate Resistance in Weeds

The herbicide glyphosate is more environmentally benign than the herbicides that it replaces. Glyphosate controls a wide array of weeds and is used on most of the HT corn, soybeans, and cotton grown in the United States. Glyphosate has been the most heavily used pesticide in the United States since 2001 (Grube et al., 2011), due in part to the popularity of HT crops and the steady decline in its price following the expiration of glyphosate's patent in 2000 (Duke and Powles, 2008).[31] Because the pollen and seeds of many different weed species can disperse between farms in the atmosphere and in conjunction with the movement of animals and farm equipment, economic incentives for adopting best management practices (BMPs) that maintain the effectiveness of glyphosate over time are reduced (Miranowski and Carlson, 1986).[32] The economic and biological impacts associated with any farmer's pesticide-use decisions will accrue not only to that farmer, but to other nearby farmers as well. Unless resistance management is coordinated across farms, economic incentives for farmers to account for the effects of their decisions on resistance are reduced, even on their own farms. This is because the effectiveness and longrun economic benefits of using BMPs to manage resistance depend on the level of adoption by nearby farmers, while the short-run costs of BMP adoption are borne solely by the adopters.[33] In this setting, resistance can evolve at an economically inefficient rate because market-based economic incentives are insufficient to promote an efficient level of BMP adoption (Hueth and Regev, 1974; Feder and Regev, 1975).

This reduction in economic incentives to adopt BMPs and the economic and environmental benefits associated with the HT crop-glyphosate combination have contributed to an overreliance on glyphosate and a concomitant reduction in the diversity of weed management practices by U.S. crop producers. This, in turn, has contributed to the evolution of glyphosate resistance in some weed species and a shift in weed composition in fields, favoring weeds that are naturally resistant to glyphosate. This leads to higher management costs, reduced yields and profits, and increased use of less environmentally benign herbicides. Glyphosate resistance is currently documented in 14 U.S. weed species (Heap, 2012), and the potential exists for much more acreage to be affected (Frisvold et al., 2009; Shaw et al., 2011).[34]

Because no new major herbicide chemistry has been made commercially available in the last 20 years, and because few new ones are expected to be available soon (Harker et al., 2012), many plant scientists believe that slowing the rate of glyphosate resistance and the spread of glyphosate-resistant (GR) weeds are among the most important problems facing U.S. crop producers (NRC, 2010, 2012). In addition, private and public programs seeking to promote the adoption of BMPs are in their infancy and do not address the reduced incentives to adopt BMPs caused by the ability of weed seeds to disperse between farms—the programs do not discourage the use of weed management practices that contribute to resistance.

Best management practices (BMP) may help sustain the efficacy of HT crops. Because weeds tend to inherit resistance to glyphosate as a dominant trait, the mandatory refuge requirement, which has been successful in sustaining the efficacy of Bt crops, might not be a viable option for HT crops (NRC, 2010). Depending on the weed, several BMPs, which are relatively difficult to monitor and enforce, might be required. These include using at least one other herbicide (particularly a residual herbicide that takes longer to decompose and thus stays in the soil longer), rotating crops, increasing the intensity of tillage, cleaning equipment between use in different fields to prevent the spread of weed seeds and pollen, and optimizing application by using the application rate recommended on the herbicide label and applying herbicides at the appropriate time and uniformly throughout the field. Some of these practices have been associated with increased weed management costs (Hurley et al., 2009), and many farmers, perhaps due partly to the incentive problems described above, are only adopting BMPs in the presence of glyphosate-resistant weeds, as opposed to adopting preventative approaches.[35]

Another approach currently being promoted by technology providers is the use of HT crops that are tolerant to two herbicides. However, the commercial availability of these types of crops does not address the incentive problem caused by the ability of weeds to disperse between farms. At least one HT crop provider is issuing rebates to growers who plant specific HT crop varieties, use glyphosate herbicides manufactured by that HT crop technology provider, and agree to use pre-emergent, residual herbicides.[36] The rebate program promotes the use of glyphosate in combination with other herbicides, which mitigates resistance; however, the program does not fully address the reduced incentive to adopt BMPs caused by the ability of weed seeds and pollen to disperse between farms. USDA's NRCS recently initiated the Integrated Pest Management (IPM) Herbicide Resistance Weed Conservation Plan, which specifies guidelines for monitoring, recordkeeping, IPM, and conservation that satisfy criteria for soil, water, and air quality. Under the program, USDA pays farmers 75 percent of the cost of developing activity plans, which contain the minimum components needed to apply for cost-sharing assistance under the Environmental Quality Incentives Program (EQIP). This program can help promote the adoption of BMPs. However, in the absence of widespread adoption of BMPs, farmer participation might be insufficient to manage the evolution of glyphosate resistance in a manner that is optimal for crop producers.

Consumer Demand for GE Products

The successful marketing of crops produced via genetic engineering is contingent on consumer acceptance of these products (or products containing GE ingredients). Some consumers, including those in the European Union, have indicated a reluctance to consume GE products. In other countries, including the United States, expression of consumer concern is less widespread. Researchers studying markets in high-income nations often find that consumers are willing to pay a premium for non-GE products,[37] but recent studies have found that some consumers in developing countries, and others interested in second-generation traits like enhanced nutrition content, are more willing to consider GE foods. Information and types of GE technology may also affect consumer response to GE foods. In some countries, retailers have developed particular policies for GE ingredients in the foods they sell under their own brand names.

Willingness-to-Pay for GE and Non-GE Foods

Researchers have used a variety of methodologies to determine how much consumers are willing to pay for GE foods and how much they are willing to pay to avoid them (table 9). Some studies use the contingent valuation method, in which consumers are asked how much they would pay for non-GE foods. Other studies use experimental auctions, in which participants bid with actual money. Consumers' responses on a survey, however, may differ from what they are actually willing to spend while shopping (Lusk, 2003). Meta-analyses of surveys indicate that when consumers are asked how much they value a good hypothetically, the values differ from what they actually will pay in a market setting, with the size of the difference dependent on such factors as whether consumers are asked how much they are willing to accept or willing to pay, the magnitude of the hypothetical price, the type of auction used or choices offered, and the type of good being evaluated (Murphy et al., 2005; List and Gallet, 2001). Murphy et al. (2005) found that models where respondents were asked to choose among alternatives, as opposed to developing their own, were associated with less hypothetical bias. On average, consumers tend to overstate what they would pay for goods, although in a significant minority of cases, they understated what they would pay. The willingness-to-pay values therefore may only approximate what consumers will actually pay. Mather et al. (2011), combining surveys with market methodology, found that when consumers in five EU countries plus New Zealand were surveyed, they selected organic over conventional or GE fruit.[38] However, when actual fruit stalls were set up offering three different varieties of fruit, consumers in Sweden, New Zealand, and Germany bought more of the GE varieties, also labeled "spray-free," but only when they were offered at a 15-percent discount.

Consumer Acceptance of GE Foods in High-and Low-Income Countries

Research on consumer acceptance of GE foods in high-income countries such as the United States, UK, and Canada finds that consumers are willing to pay a premium for non-GE foods (Bernard and Bernard, 2010; Huffman, 2010; Hartl and Herrmann, 2009; Volinskiy et al., 2009; Bukenya and Wright, 2007; Moon et al. 2007; Huffman et al., 2007; Carlsson et al., 2007; Tegene et al., 2003; Loureiro and Hine, 2002; Burton et al., 2001; Lusk et al., 2003;

Moon and Balasubramanian, 2001). Other research studies have identified concerns about GE foods (Bernard et al., 2007; Komirenko et al., 2010). Lusk et al. (2005) found that much of the variation in premia for non-GE foods across studies can be explained by a number of factors, including whether the study was done in Europe, whether the research surveyed shoppers, whether the survey took place in person, whether the consumers were asked to give hypothetical values for willingness-to-pay, whether they were asked for values for GE or non-GE foods, what type of product was considered, and whether consumers were told the product would provide them with a direct benefit.

Table 9. Studies in which consumers were willing to pay a premium for non-GE food

Country	Good	Study	Willingness to pay premium [1]
United States	Potatoes and corn	Bernard and Bernard, 2010	In experimental nth price auction, found positive premium for non-GE food
United States	Various	Huffman, 2010	In experimental nth price auction, found 15-percent discount for GE food, but difference was only statistically significant for one of three foods
United States	Tomatoes	Bukenya and Wright,2007	Surveyed Alabama consumers willing to pay a $0.39 or 19-21 percent premium for non-GE tomatoes
United States	Vegetable oil, tortilla chips, and potatoes	Huffman et al., 2007	Found consumers willing to pay 14 percent less for GE foods
United States	Vegetable oil	Tegene et al., 2003	In experimental auctions, consumers willing to pay 14 percent more for non-GE food
United States	Potatoes	Loureiro and Hine, 2002	Customers willing to pay 5 percent more for non-GE food
United Kingdom	All foods	Burton et al.,2001	Customers indicated willingness to increase food budgets by 26-129 percent to avoid GE foods

Country	Good	Study	Willingness to pay premium [1]
United States, France, Germany, United Kingdom	Beef fed with GE feed	Lusk et al., 2003	U.S. consumers willing to pay \$2.83 and \$3.31 per lb. to avoid biotech; European consumers \$4.86 to \$11.01
United States, United Kingdom	Breakfast cereal	Moon and Balasubra-manian, 2001	Found 56 percent of UK cons-umers willing to pay a premium to avoid GE food, compared to 37 percent of U.S. consumers.
United Kingdom	Various	Moon et al., 2007	Found that consumers were willing to pay a 20-per-cent premium for non-GE products and willing to accept a discount of 23 percent for GE foods
UK, Belgium, France, Germany, New Zealand, Sweden	Fruit	Mather et al., 2011	Found that surveyed consumers offered organic, conventional, or GM fruit stated that they wanted organic, but the same consumers at roadside stalls bought GM (labeled spray-free and offered at a 15-percent discount) 15-43 percent of the time
Germany	Canola	Hartl and Herrmann, 2009	In an online survey, found that the GE version must be discounted by over 100 percent
Romania	Potatoes, Sunflower oil	Curtis and Moeltner, 2007	Found that so few of surveyed Romanians were will-ing to purchase GE foods that a premium could not be calculated
Sweden	Beef, chicken	Carlsson et al., 2007	Found that consumers were willing to pay 30 SEK/kg extra for chicken and 32.5 SEK/kg for beef fed feed not produced using GE ingredients
Norway, United States, Japan, Taiwan	Vegetable oil	Chern et al., 2002	For non-GE vegetable oil, Norwegian students were willing to pay \$1.51 (55-69 percent premium) per liter. U.S. students were willing to pay \$1.13 (50-62 percent premium), Japanese students were willing to pay \$0.88

Table 9. (Continued)

Country	Good	Study	Willingness to pay premium [1]
			(33-40 percent premium), and Taiwanese students were willing to pay $0.45 (17-21 percent premium)
Norway	Bread	Grimsrud, et al., 2004	Consumers required discounts of 37 to 63 percent to buy GE bread; one-fourth were willing to buy with no discount
Australia	Beer	Burton and Pearse, 2002	Younger Australian consumers would pay $A 0.72 less and older consumers $A 0.40 less for beer made with GE barley
Canada	Canola	Volinskiy et al, 2009	In a shopping experiment, found that consumers would pay Canadian $0.45 (20-30 percent) premium for non-GE canola
Canada		West et al., 2002	83 percent of consumers ascribed a lower value to several GE food products
China	Vegetable oil	Hu et al., 2006	Consumers would consume GE product with a 14-percent discount after hearing basic or positive info-rmation, and a 66-percent discount after hearing negative information
China	Soybean oil and rice	Lin et al., 2006	Consumers on average would pay a 52-percent premium for non-biotech foods
France	Biscuits	Noussair, et al., 2004	35 percent of consumers were unwilling to purchase GE foods, and 42 percent were willing to purchase them if they were less expensive
United States	Various	Rousu et al., 2004	Consumers reduced their demand by an average of 7-13 percent for each food product having 1-perc-ent and 5-percent tolerance levels for GE material relative to food not produced using GE ingredients

[1] Lusk et al., 2005 contains a more exhaustive review of the literature prior to 2005.

[2] Across all information treatments.

More recent research on consumer willingness-to-pay for GE foods has focused on consumers in developing countries and has yielded different results than in wealthier nations (table 10). Several authors found that consumers are willing to pay a slight premium for GE foods in India (Krishna and Qaim, 2008; Anand et al., 2007), Kenya (Kimenju and De Groote, 2008), and China (Li et al., 2002). The few studies that have considered second generation attributes like nutrition have also found willingness-to-pay a premium for GE foods in India and Brazil (Anand et al., 2007; Gonzalez et al., 2009a). However, these findings are not universal across all developing countries. Hu et al. (2006) found that, on average, consumers in Nanjing would consume GE vegetable oil at a discount of 14 percent if presented with basic information or positive information and would require a discount of 66 percent if presented with negative information. Lin et al. (2006) also found that Chinese consumers would pay an average premium of 52 percent for non-GE foods. In addition, in Romania, a lower income country that is also in the EU, Curtis and Moeltner (2007) could not calculate a premium for non-GE goods over GE goods since too few in their survey of Romanians were willing to purchase GE goods. A cluster analysis of Brazilian stakeholders in the debate over GE foods found that while many respondents perceived little or no risk from GE foods, some were skeptical of the benefits (Gonzalez et al., 2009b).

Some studies found that consumers, on average, would pay a premium for the non-GE version of the product while some would be willing to purchase GE foods without a premium (Lin et al., 2006). Bukenya and Wright (2007) found that younger consumers were willing to pay a premium for GE versions of the product.

More research is beginning to focus on second-generation attributes. Many of the currently marketed varieties of GE foods come from crops that have been engineered to decrease yield losses to pests and/or reduce costs of production (first generation). Second-generation attributes refer to genetically engineered characteristics of the foods themselves, such as extra vitamins that might make the food more attractive to consumers. Lusk et al. (2005) examined the literature up until 2005 and found that benefits to the consumer were significant in explaining the size of the premium consumers would pay for a non-GE food. Huffman (2010) found that U.S. consumers were willing to pay a premium for vitamin-enhanced GE food, as did Lusk (2003). Gonzalez et al. (2009) and Anand et al. (2007) found the same thing for consumers in Brazil and India. With an online survey of German consumers, Hartl and Herrmann (2009) found that GE enhancement of the Omega-3 content of foods or the addition of cholesterol-fighting compounds reduced the discount that GE foods had to offer relative to non-GE products. Boccaletti and Moro (2000) found that a sample of Italian consumers were willing to pay more for GE foods with improved nutritional qualities and lower pesticide use.

Table 10. Studies in which consumers were willing to pay a premiumfor genetically engineered (GE) food or GE food with enhanced characteristics

Country	Good	Study	Willingness to pay premium [1]
United States	Good	Huffman, 2010	Found that consumers would pay a 19- to 26-percent [2] premium for a product with intragenic addition of vitamins over a plain labeled product
United States	Golden Rice	Lusk, 2003	Customers willing to pay $0.93 for GE "golden rice" with added vitamin A, $0.65-0.75 for regular rice
Germany	Canola	Hartl and Herrmann, 2009	In an online survey, also found that consumers were willing to pay 1.37 Euros/ half-liter extra for GE oil with Omega3's and 0.80 euros per half-liter for GE oil with cholesterol-reducing compounds, which reduced but didn't eliminate the GE discount
Italy		Bocatelli and Moro, 2001	Consumers willing to pay a positive amount for GE attributes; 66 percent did not require a premium to consume GE foods
China	Rice	Li et al., 2002	80 percent of consumers did not require a premium to purchase GE rice and on average were willing to pay a 38-percent premium for GE rice and a 16-per- cent premium for GE soy oil
Brazil	Vitamin A fortified cassava	Gonzalez et al., 2009	Found surveyed consumers willing to pay 64-70 percent more for GE Vitamin-A fortified cassava
India	Bt veg- etables	Krishna and Qaim, 2008	Found surveyed consumers willing to pay 1.5 percent premium for GE Bt (pest resistant) vegetables
India	Wheat	Anand et al., 2007	Found that if given no info, consumers will pay a 7-percent premium for GE foods; positive info leads to a 10-percent

Country	Good	Study	Willingness to pay premium [1]
			premium, negative info leads to a negative 139-percent premium (discount) for GE foods, and positive info on heart-healthy characteristics leads to a 23-percent premium for GE foods
Kenya	Maize meal	Kimenju and De Groote, 2008	Consumers surveyed in 2003 would pay a 13.8-percent premium for GE food

[1] Lusk et al., 2005 contains a more exhaustive review of the literature prior to 2005.
[2] Across all information treatments.

A new area of research has been the contrast between intragenic and transgenic goods. Intragenic goods are created by transferring genes from a plant of the same species, but of a different variety, as opposed to transferring a gene from another species or type of plant. Huffman (2010), using experimental auctions, found that consumers discounted GE foods, but were willing to pay a premium for intragenic foods that had enhanced vitamin content versus a plain-labeled product. The difference between the premia for intragenicand transgenic-enhanced vitamins, however, was not statistically significant unless pro-biotech information was given to consumers. A survey of stakeholders in the potato industry (Toevs et al., 2011) found that certain categories of stakeholders (women, Canadians) were optimistic about intragenic potato varieties. More research remains to be done to determine whether consumers as a whole will find intragenic foods more acceptable than transgenic foods.

Effect of Information on the Desire of Consumers to Purchase GE Foods

Several studies have also considered the impact of information on the desire of consumers to purchase GE foods, and the results have varied. Huffman (2010), Huffman et al. (2007), Hu et al. (2006), and Tegene et al. (2003) found that positive information regarding biotechnology increased the willingness-to-pay for GE foods, while negative information reduced it. Onyango et al. (2004) found that those given both positive and negative information were less willing to buy GE foods than those given only positive information. Martinez-Poveda et al. (2009) found that previous knowledge of

GE technology reduces the effects of negative information on the perception of GE foods, but could increase concern for health. Boccaletti and Moro (2000) found that previous knowledge increased the willingness-to-pay for positive GE attributes, while Lusk (2003) found that lack of previous knowledge increased willingness-to-pay.

Other studies have found that consumers value certain types of information. Hu et al. (2009) argued that those choosing to access information about GE foods may be different types of consumers than those who don't access information about GE foods. They found that consumers who voluntarily access *general* information on GE foods are more likely to buy them, while those who access *environmental* information related to GE foods are less likely to buy them. Rousu and Lusk (2009) found that providing consumers with information on the environment was more likely to change consumer purchasing behavior with respect to GE foods, while information on the beneficial impact of GE foods in developing countries created more value for the consumer.

Evidence from Retail Settings

Market settings offer examples of retailers' efforts to consider consumer preferences for GE foods. Some retailers do not have policies that explicitly address GE foods. Other retailers, mostly in the EU, have explicit policies stating that GE ingredients will not be used in their brand name food products.[39] Some companies have even introduced lines of meat and eggs from animals not fed on GE feed (Agriculture and Agri-food Canada, 2007; ASDA, 2011; Carrefour, 2011b; ECCC, 2008; Tesco, 2011; SHAFE, 2011).

Lusk et al. (2005) found that consumers in Europe were willing to pay more for non-GE foods than consumers in other regions. Thus, we might expect to see more responsiveness on the part of European retailers, and indeed some of them have developed auditing procedures for their suppliers (Tesco, 2011). Store visits by researchers in 10 EU countries found few (between 1 and 27) products with GE ingredients in grocery stores in 7 of the countries (King's College, 2008). The results of Mather et al. (2011) suggest that there may be circumstances under which consumers in a few EU countries would purchase GE foods.

A market exists for non-GE products in the United States, as some U.S. retailers do offer non-GE products, and U.S. consumers wishing to avoid GE ingredients may also purchase organic products. However, the share of this market in the United States is still small compared to the widespread marketing of non-GE goods in the EU. For example, the four largest retail chains in the UK[40] all indicate on their websites that their own-brand products do not include biotech ingredients (Tesco website, 2011; ASDA website, 2011; Sainsbury's website, 2013; Wm Morrison website, 2013). They also have tried developing brands of meat from animals fed non-GE feed, but some of the chains have been unable to source enough feed to maintain production (Tesco Food News, 2013; Wm Morrison website, 2013). In contrast, of the four largest U.S. grocery retail chains,[41] two make no mention of GE foods on their websites or corporate responsibility reports, one indicates that non-GE ingredients are not yet defined, and one will make one of its inhouse product lines non-GE in the coming year (Walmart, 2013a; Walmart, 2013b; Kroger, 2013a, Kroger, 2013b; Publix website, 2013; Safeway/Vons website, 2013). Thus, U.S. supermarkets do not perceive the same advantage from marketing non-GE goods that the UK retailers do.

Further evidence comes from new product introductions in the United States. Of the 7,637 new food or food supplement products introduced between February 12, 2010, and February 11, 2011, as documented by the Datamonitor database, 2.6 percent advertise[42] that they do not include GE ingredients, 8 percent advertise that they are organic, and another 2.8 percent indicate that they at least have some organic or non-GE ingredients (Datamonitor, 2010-2011). Organic acreage of corn and soy, two potential sources of verified non-GE ingredients for U.S. food producers, remain a small share of the total acreage, with organic soy constituting 0.17 percent of total U.S. production and organic corn constituting 0.26 percent of total U.S. production in 2011 (USDA-ERS, 2013). Whether patterns of consumer approval have changed over time is not clear. International Food Information Council (FIC) polls seem to indicate that the percentage with favorable opinions of GE foods in the United States fell between 2003 and 2008, but it has recently risen somewhat. In terms of the more rigorous studies cited in this report, even in the United States and in the United Kingdom, for which we cite several studies from different time periods, the temporal patterns are not clear enough to draw definite conclusions.

CONCLUSION

A large majority of U.S. farmers have adopted GE seeds for corn, soybeans, and cotton since their commercial introduction over 15 years ago.

Despite the higher prices of GE seeds compared to conventional seed, farmers realize economic benefits from growing GE crops through higher crop yields, and/or lower pesticide costs, and management time savings.

Farmers will continue to use GE seeds as long as these seeds benefit them. However, it is not clear that first-generation GE seeds will benefit farmers indefinitely. With the help of refuges, the emergence of insect resistance to Bt crops has been low and of little economic significance over the first 15 years, but there are some indications that insect resistance is developing to some Bt traits in some areas and resistance to the herbicide glyphosate has already evolved in certain weed populations. Best management practices can help delay the evolution of resistance and sustain the efficacy of HT crops.

An important issue beyond the scope of this report is the coexistence of crop production systems. According to the USDA Advisory Committee on Biotechnology and 21st Century Agriculture— AC21 (2012), coexistence is defined as the "concurrent cultivation of crops produced through diverse agricultural systems including traditionally produced, organic, identity preserved, and genetically engineered crops." USDA supports all these crop production systems and wants each to be "as successful as possible providing products to markets in the United States and abroad."[43] ERS is collecting data and conducting a study on several aspects of the economics of coexistence of organic, non-GE, and GE crops.

REFERENCES

Agriculture and Agri-food Canada. (1996). *Exporting Processed Foods Containing GM Ingredients to Europe; Final Report Prepared for the Ontario Ministry of Agriculture*, Food & Rural Affairs, May 16.

Anand, A., Mittelhammer, R., McCluskey, J. J. & Jill, J. (2007). "Consumer Response to Information and Second-Generation Genetically Modified Food in India" *Journal of Agricultural and Food Industrial Organization*, *5*(1). http://www.degruyter.com/view/j/jafio.2007.5.1/jafio.2007.5.1.1138/jafio.2007.5.1.1138.xml?format=INT

Andow, D., Hilbeck, A. & Van Tuat, N. (2008). "Challenges and Opportunities with Bt Cotton in Vietnam," *Environmental Risk Assessment of Genetically Modified Organisms*. Volume *4*, Chapter 13.

Arnold, J. C., Shaw, D. R. & Medlin, C. R. (1998). "Roundup Ready TM Programs Versus Conventional Programs: Efficacy, Varietal Performance, and Economics." *Proceedings of the Southern Weed Science Society, Southern Weed Science Society*. (*51*), 272-73.

ASDA Website. (2011). http://your.asda.com/2011/2/21/giving-you-a-healthier-choice

Baute, T. S., Sears, M. K. & Schaafsma, A. W. (2002). "Use of Transgenic *Bacillus thuringiensis* Berliner Corn Hybrids to Determine the Direct Economic Impact of the European Corn Borer (*Lepidoptera: Crambidae*) on Field Corn in Eastern Canada. *Journal of Economic Entomology*. *95*(1), 57-64.

Belson, N. A. (2000). "U.S. Regulation of Agricultural Biotechnology: An Overview. *AgBioForum, 3*(4), 268-280. http://www.agbioforum.org

Bernard, J. C. & Bernard, D. J. (2010. "Comparing Parts with the Whole: Willingness to Pay for Pesticide-Free, Non-GM, and Organic Potatoes and Sweet Corn," *Journal of Agricultural and Resource Economics, 35*(3), (December), 457-75.

Bernard, J. C., Pesek, J. D. Jr. & Xiqian, Pan. (2007). "Consumer Likelihood to Purchase Chickens with Novel Production Attributes," *Journal of Agricultural and Applied Economics, 39*(3), December), 581-96

Bernard, J.C., Pesek, J. D. Jr. & Fan, C. (2004). Performance results and characteristics of adopters of genetically engineered soybeans in Delaware. *Agricultural and Resource Economics Review, 33*(2), 282-292.

Boccaletti, S. & Moro, D. (2000). "Consumer Willingness-To-Pay For GM Food Products In Italy," *AgBioForum, 3*(4), 259-267. http://www.agbioforum.org

Bradley, K. W., Hagood, E. S. Jr. & Davis P. H. (2004). Trumpetcreeper (*Campsis radicans*) control in double-crop glyphosate-resistant soybean with glyphosate and conventional herbicide systems. *Weed Technology*, (*18*), 298–303.

Bryant, K. J., Robertson, W. C. & Lorenz, G. M. III. (1998). "Economic Evaluation of Bollgard Cotton in Arkansas." In *Proceedings of the Beltwide Cotton Conferences*, National Cotton Council, Memphis, TN. (*1*), 388-89.

Bukenya, J. O. & Wright, N. R. (2007). "Determinants of Consumer Attitudes and Purchase Intentions with Regard to Genetically Modified Tomatoes," *Agribusiness, 23*(1), (Winter), 117-30

Bullock, D. S. & Nitsi, E. I. (2001). Roundup ready soybean technology and farm production costs, measuring the incentive to adopt genetically modified seeds. *American Behavioral Scientist, 44*(8), 1283-1301.

Burton, M., Rigby, D., Young, T. & James, S. (2001). "Consumer Attitudes to Genetically Modified Organisms in Food in the UK," *European Review of Agricultural Economics, 28*(4), (December), 479-98.

Burton, M. & Pearse, D. (2002). "Consumer Attitudes Towards Genetic Modification, Functional Foods & Microorganisms: A Choice Modeling Experiment for Beer," *AgBioForum, 5*(2), 51-58. http://www.agbioforum.org

Carpenter, J. & Gianessi, L. (1999). Herbicide Tolerant Soybeans: Why Growers are Adopting Roundup Ready Varieties. *AgBioForum, 2*(2), (Spring), 65-72. http://www.agbioforum.org

Carrefour Website. (2011a). http://www.carrefour.com/cdc/responsible-commerce/our-commitment-to-the-environment/responsible-sourcing/

Carrefour Website. (2011bopment-news/france-carrefour_s-_reared-without-gmos_-labelling.html.

Carlsson, F., Frykblom, P. & Lagerkvist, C. J. (2007). "Consumer Benefits of Labels and Bans on GM Foods—Choice Experiments with Swedish Consumers," *American Journal of Agricultural Economics, 89*(1), 152-161.

Carrière, Y., Ellers-Kirk, C., Sisterson, M., Antilla, L., Whitlow, M., Dennehy, T. J. & Tabashnik, B. E. (2003). Long-term regional suppression of pink bollworm by Bacillus thuringiensis cotton. *Proceedings of the National Academy of Sciences, (100)*, 1519–1523.

Cattaneo, M. G., Yafuso, C. M., Schmidt, C., Huang, C.-Y., Rahman, M., Olson, C., Ellers-Kirk, C., Orr, B.J., Marsh, S. E., Antilla, L., Dutilleul, P. & Carrière, Y. (2006). "Farm-scale evaluation of the impacts of transgenic cotton on biodiversity, pesticide use and yield," *Proceedings of the National Academy of Sciences of the United States of America, 103*(20), 7571–7576.

Chern, W. S., Rickertsen, K., Tsuboi, N. & Fu, T. (2002). "Consumer Acceptance and Willingness to Pay for Genetically Modified Vegetable Oil and Salmon: A Multiple-Country Assessment," *AgBioForum, 5*(3), 105-112.

Cox, W. J., Hanchar, J. & Shields, E. (2009). "Stacked Corn Hybrids Show Inconsistent Yield and Economic Responses in New York," *Agronomy Journal, 101*(6), 1530–1537.

Culpepper, A. S. & York, A. C. (1998). "Weed Management in Glyphosate-Tolerant Cotton," *The Journal of Cotton Science, (4)*, 174-185.

Curtis, K. R. & Moeltner, K. (2007). "The Effect of Consumer Risk Perceptions on the Propensity to Purchase Genetically Modified Foods in Romania," *Agribusiness*, *23*(2), (Spring), 263-78.

Datamonitor. (2010-11). *Product Launch Analytics,* (formerly Productscan). Database.

Dauer, J. T., Luschei, E. C. & Mortensen, D. A. (2009). "Effects of Landscape Composition on Spread of an Herbicide-Resistant Weed," *Landscape Ecology*, *24*, 735-747.

Delannay, X., Bauman, T. T., Beighley, D. H., Buettner, M. J., Coble, H. D., DeFelice, M. S., Derting, C. W., Diedrick, T. J., Griffin, J. L., Hagood, E. S., Hancock, F. G., Hart, S. E., LaVallee, B. J., Loux, M. M., Lueschen, W. E., Matson, K. W., Moots, C. K., Murdock, E., Nickell, A. D., Owen, M. D. K., Paschall E. H. II, Prochaska, L. M., Raymond, P. J., Reynolds, D. B., Rhodes, W. K., Roeth, F .W., Sprankle, P. L., Tarochionc, L. J., Tinius, C. N., Walker, R. H., Wax, L. M., Weigelt, H. D. & Padgette, S. R. (1995). "Yield Evaluation of a Glyphosate Tolerant Soybean Line after Treatment with Glyphosate," *Crop Science*, (*35*), 1461-67.

Dillehay, B. L., Roth, G. W., Calvin, D. D., Karatochvil, R. J., Kuldau, G. A., & Hyde, J. A. (2004). "Performance of Bt Corn Hybrids, their Near Isolines, & Leading Corn Hybrids in Pennsylvania and Maryland," *Agronomy Journal*, (*96*), 818-824.

Duffy, M. (2001). "Who Benefits from Biotechnology?" Paper presented at the American Seed Trade Association meeting. Chicago, IL. Dec. 5-7.

Duke, S. O. & Powles, S. B. (2008). "Glyphosate: A Once-In-A-Century Herbicide," *Pest Management Science*, *64*(4), 319–325.

El-Osta, H. & Morehart, M. (1999). "Technology Adoption Decisions in Dairy Production and the Role of Herd Expansion," *Agricultural and Resource Economics Review*, *28*(1), 84-95.

European Community of Consumer Cooperatives (ECCC). (2008. "Consumer Cooperative Social Responsibility: Bringing Values into Retail Business," Dec. http://www.eurocoop.org/dmdocuments/publication_report/CCSR publication08.pdf

European Food Safety Authority (EFSA). (2012. "Final review of the Séralini et al. (2012a) publication on a 2-year rodent feeding study with glyphosate formulations and GM maize NK603 as published online on 19 September 2012 in Food and Chemical Toxicology," *EFSA Journal*, *10*(11), 2986.

Falck-Zepeda, J. B., Traxler, G. & Nelson, R. G. (2000). "Rent creation and distribution from biotechnology innovations: The case of Bt cotton and herbicide-tolerant soybeans in 1997," *Agribusiness*, *16*(1), 21–32.

Fang, M., Motavalli, P. P., Kremer, R. J. & Nelson, K. A. (2007). "Assessing changes in soil microbial communities and carbon mineralization in Bt and non-Bt corn residue-amended soils." *Applied Soil Ecology*, *37*(1&2), (Oct.), 150-160.

FDA (Food and Drug Administration). (2005). Center for Food Safety and Applied Nutrition. *Biotechnology.* http://www.cfsan.fda.gov/~lrd/biotechm.html. May.

Federal Register. (2011). "*Monsanto Co., Determination of Nonregulated Status of Corn Genetically Engineered for Drought Tolerance*." Volume *76*, Issue 248, Dec. 27.

Feder, G. & Regev. U. (1975). "Biological Interactions and Environmental Effects in the Economics of Pest Control," *Journal of Environmental Economics and Management*, (*2*), 75-91.

Fernandez-Cornejo, J. (2004). *The Seed Industry in U.S. Agriculture: An Exploration of Data and Information on Crop Seed Markets, Regulation, Industry Structure and Research and Development.* AIB-786, Economic Research Service, U.S. Department of Agriculture, Feb.

Fernandez-Cornejo, J. & Jiayi, Li. (2005). "The Impacts of Adopting Genetically Engineered Crops in the USA: The Case of Bt Corn." Selected Paper presented at the American Agricultural Economics Association meetings. Providence, RI. http://ageconsearch.umn.edu/bitstream/19318/1/sp05fe01.pdf

Fernandez-Cornejo, J. & Caswell, M. (2006). *The First Decade of Genetically Engineered Crops in the United States.* EIB-11, Economic Research Service, U.S. Department of Agriculture, April.

Fernandez-Cornejo, J. & Wechsler, S. (2012). "Revisiting the Impact of Bt Corn Adoption by U.S. Farmers," *Agricultural and Resource Economics Review*, *41*(3), (Dec.), 377-390.

Fernandez-Cornejo, J. & McBride, W. D. (2002). *Adoption of Bioengineered Crops.* AER-810, Economic Research Service, U.S. Department of Agriculture, May.

Fernandez-Cornejo, J., Hendricks, C. & Mishra, A. (2005). "Technology Adoption and Off-farm Household Income: The Case of Herbicide-Tolerant Soybeans," *Journal of Agricultural and Applied Economics*, *37*(3), (Dec.), 549-563.

Fernandez-Cornejo, J., with contributions from Mishra, A., Nehring, R., Hendricks, C., Gregory, A. & Southern, M. (2007). *Off-Farm Income, Technology Adoption and Farm Economic Performance.* ERR-36, Economic Research Service, U.S. Department of Agriculture, Feb.

Fernandez-Cornejo, J., Klotz-Ingram, C. & Jans, S. (2000). "Farm-Level Effects of Adopting Genetically Engineered Crops in the U.S.A.," *Transitions in Agrobiotech: Economics of Strategy and Policy*, W.L. Lesser, ed. Food Marketing Research Center, University of Connecticut and Department of Research Economics, University of Massachusetts.

Fernandez-Cornejo, J., Klotz-Ingram, C. & Jans, S. (2002). "Farm-Level Effects of Adopting Herbicide-Tolerant Soybeans in the U.S.A.," *Journal of Agricultural and Applied Economics*, *34*(1), 149-163.

Fernandez-Cornejo, J., Klotz-Ingram, C., Heimlich, R., Soule, M., McBride, W. & Jans, S. (2003). "Economic and Environmental Impacts of Herbicide-Tolerant and Bt Crops in the United States," *The Economic and Environmental Impacts Agbiotech: A Global Perspective,* Nicholas Kalaitzandonakes, ed. New YorkY: Kleuwer Academic/Plenum Publishers.

Fernandez-Cornejo, J., Hallahan, C., Nehring, R., Wechsler, S. & Grube, A. (2012). "Conservation Tillage, Herbicide Use, & Genetically Engineered Corps in the U.S.A.: The Case of Soybeans." *AgBioForum*, *15*(3), (Dec.), 231-241. http://www.agbioforum.com.

Ferrell, J. A. & Witt, W. W. (2002). "Comparison of glyphosate with other herbicides for weed control in corn (*Zea mays*): efficacy and economics,." *Weed Technology*, (*16*), 701-706.

Fischer, R. A. & Edmeades, G. O. (2010). "Breeding and Cereal Yield Progress**,"** *Crop Science*, *50*(March-April), S-85-S-98.

Food and Agriculture Organization of the United Nations (FAO). (2001). *Glossary of Biotechnology for Food and Agriculture A Revised and Augmented Edition of the Glossary of Biotechnology and Genetic Engineering.* FAO Research and Technology paper 9, (A. Zaid, H.G. Hughes**,** E. Porceddu, and F. Nicholas), Rome. http://www.fao.org/docrep/004/y2775e/y2775e00.htm

Frisvold, G. B., Hurley, T. M. & Mitchell, P. D. (2009). "Overview: Herbicide Resistant Crops– Diffusion, Benefits, Pricing, and Resistance Management," *AgBioForum*, *12*(3&4), 244-248. http://www.agbioforum.org

Gardner, J. G., Nehring, R. F. & Nelson, C. H. (2009). Genetically modified crops and household labor savings in U.S. crop production," *AgBioForum, 12*(3&4), 303-312. http://www.agbioforum.org.

Gassmann, A. J., Petzold-Maxwell, J. L., Keweshan, R. S. & Dunbar, M. W. (2011). "Field-Evolved Resistance to Bt Maize by Western Corn Rootworm," *PLOS ONE*, 6(7), e22629. doi:10.1371/journal.pone.

0022629. July 29. http://www.plosone.org/article/info%3Adoi%2F10.1371%2Fjournal.pone.0022629

Geissler, S. (2009). "CONAD Prospers with Multi-Tier PL Program." *Private Label*, Nov./Dec.

Gibson, J. W. IV, Laughlin, D., Lutrell, R. G., Parker, D., Reed, J. & Harris, A. (1997). "Comparison of Costs and Returns Associated with *Heliothis* Resistant Bt Cotton to Non-Resistant Varieties," *Proceedings of the Beltwide Cotton Conferences*. Cotton Weed Science Research Conference.

Golan, E., Kuchler, F., Mitchell, L., Greene, C. & Jessup, A. (2001). *Economics of Food Labeling*. U.S. Department of Agriculture, Economic Research Service. AER-793. Jan.

Goldman, D. L., Guy, C., McCelland, M. & Kendig, A. (1996). "Weed Control Systems in Roundup Ready® Cotton," *Proceedings of the Beltwide Cotton Conferences*. Cotton Weed Science Research Conference.

Gonzalez, C., Johnson, N. & Qaim, M. (2009). "Consumer Acceptance of Second-Generation GM Foods: The Case of Biofortified Cassava in the North-East of Brazil," *Journal of Agricultural Economics*, *60*(3), (Sept.), 604-24.

Gonzalez, C., Garcia, J. & Johnson, N. (2009). "Stakehold Positions Toward GM Food: the Case of Vitamin A Fortified Cassava in Brazil," *AgBioForum*, *12*(3-4), 382-93.

Gordon, H. S. (1954). "The Economic Theory of a Common-Property Resource: The Fishery," *Journal of Political Economy*, *62*(2), 124-142.

Greene, C. & Smith, K. (2010). "Can Genetically Engineered and Organic Crops Coexist?" *Choices*, *25*(2)(2nd Quarter). http://www.choicesmagazine.org/magazine/article.php?article=131.

Greene, W. (1997). *Econometric Analysis*, 3rd edition. Upper Saddle River, NJ: Prentice-Hall.

Grimsrud, K. M., McCluskey, J., Loureiro, M. & Wahl, T. (2004). "Consumer Attitudes to Genetically Modified Foods in Norway," *Journal of Agricultural Economics*, *55*(1), (March), 75-90.

Grube, A. H., Donaldson, D., Kiely, T. & Wu, L. (2011). *Pesticides Industry Sales and Usage 2006 and 2007 Market Estimates*, U.S. Environmental Protection Agency, Feb.

Hartl, Jochen & Roland Herrmann. (2009). "Do They Always Say No? German Consumers and Second-Generation GM Foods," *Agricultural Economics*, *40*(5), (Sept.), 551-60.

Heap, I. (2012). "Glycines (G/9) Resistant Weeds by Species and Country," *International Survey of Herbicide Resistant Weeds*, http://www.weedscience.org/Summary/UspeciesMOA. asp?lstMOAID =12

Hedrick, P. W. (2000). *Genetics of Populations*. 2nd ed., Jones and Bartlett Publishers: Sudbury, MA.

Heisey, P. W. & Fuglie, K. O. (2012). "Private Research and Development for Crop Genetic Improvement, " *Research Investments and Market Structure in the Food Processing, Agricultural Input, and Biofueld Industries Worldwide*. U.S. Department of Agriculture, Economic Research Service. AER-130.

Hu, W., Zhong, F. & Ding, Y. (2006). "Actual Media Reports on GM Foods and Chinese Consumers' Willingness to Pay for GM Soybean Oil," *Journal of Agricultural and Resource Economics*, *31*(2), 376-90.

Huffman, W. E. (2010). *Consumer Acceptance of Genetically Modified Foods: Traits, Labels and Diverse Information* Iowa State University, Department of Economics, Staff General Research Papers.

Horowitz, J., Ebel, R. & Ueda, K. (2010). *No-Till Farming is a Growing Practice*. U.S. Department of Agriculture, Economic Research Service. EIB-70, Nov.

Huffman, Wallace E., Matthew Rousu, Jason, F. Shogren & Abebayehu Tegene. (2007). "The Effect of Prior Beliefs and Learning on Consumers' Acceptance of Genetically Modified Foods," *Journal of Economic Behavior and Organization*, *63*(1), (May), 193-206.

Hutchinson, W. D., Burkness, E. C., Hellmich, R. L., Kaster, L. V., Hunt, T. E., Wright, R. J., Pecinovsky, K., Rabaey, T. L., Flood, B. R., Raun, E. S., Mitchell, P. D., Moon, R. D., Leslie, T. W., Fleischer, S. J., Abrahamson, M., Hamilton, K. L., Steffey, K. L. & Gray, M. E. (2010). "Areawide Suppression of European Corn Borer with BT Maize Reaps Savings to Non-Bt Maize Growers," *Science*, (330), 222-225.

Hueth, D. & Regev. U. (1974). "Optimal Agricultural Pest Management with Increasing Pest Resistance," *American Journal of Agricultural Economics*, *56*(3), 543-552.

Hurley, T. M., Babcock, B. A. & Hellmich, R. L. (2001). "Bt Corn and Insect Resistance: An Economic Assessment of Refuges," *Journal of Agricultural and Resource Economics*, (*26*), 176-194.

Hurley, T. M., Mitchell, P. D. & Frisvold, G. B. (2009). "Weed Management Costs, Weed Best Management Practices, & the Roundup Ready® Weed Management Program," *AgBioForum*, *12*(3&4), 281-290.

Hyde, J., Martin. M. A., Preckel, P. V. & Edwards, C. R. (1999). "The Economics of Bt Corn: Valuing Protection from the European Corn Boer," *Review of Agricultural Economics*, (*21*)(2), 442-454.

Information Systems for Biotechnology (ISB). (2013). *Search Biotechnology Data*. http://www.isb.vt.edu/data.aspx

International Service for the Acquisition of Agri-Biotech Applications (ISAAA). (2012). *Global Status of Commercialized Biotech /GM Crops: 2012*. ISAAA Brief 44-2012. http://www.isaaa.org/resources/publications/briefs/44/executivesummary/default.asp

Janssen, C. & Hill, P. (1994). *What is Conservation Tillage?* Conservation Tillage Series. CT-1 Cooperative Extension Service, Purdue University, West Lafayette, IN. Nov.

Kalaitzandonakes, N., Magnier, A. & Miller, D. (2010-2011). "A Worrisome Crop? Is there Market Power in the Seed Industry? " *Regulation*. Winter: 20-26.

Keeling, W., Dotray, P., Jones, C. & Sunderland, S. (1996). "Roundup® Ready Cotton: A Potential New Weed Management Tool for the Texas High Plains," *Proceedings of the Beltwide Cotton Conferences*. Cotton Weed Science Research Conference.

Kimenju, S. C. & De Groote, H. (2008). "Consumer Willingness to Pay for Genetically Modified Food in Kenya," *Agricultural Economics*, *38*(1), (Jan.), 35-46.

King, J. & Heisey, P. (2003). "Ag Biotech Patents: Who's Doing What? " *Amber Waves*, Vol. *1*, Issue 5, U.S. Department of Agriculture, Economic Research Service, Nov.

King, J. & Heisey, P. (2004). *Agricultural Biotechnology Intellectual Property* Data Set. U.S. Department of Agriculture, Economic Research Service. http://webarchives.cdlib.org/sw1s17tt5t/http://ers.usda.gov/data/AgBiotechIP/

King's College London. (2008). "Do European Consumers Buy GM Foods? " European Commission, Framework Six, Project No. 518435, Oct. http://www.kcl.ac.uk/medicine/research/divisions/dns/projects/consumerchoice/downloadfiles/Fullreport1-11.pdf

Komirenko, Z., Veeman, M. M. & Unterschultz, J. R. (2010). "Do Canadian Consumers Have Concerns about Genetically Modified Animal Feeds? " *AgBioForu*, 13(3), 242-50.

Kott, P. S. (1998). *Using the Deleted-A-Group Jackknife Variance Estimator in NASS Surveys*. U.S. Department of Agriculture, National Agricultural Statistics Service.

Krishna, V. V. & Qaim, M. (2008). "Consumer Attitudes toward GM Food and Pesticide Residues in India," *Review of Agricultural Economics*, *30*(2) (Summer); 233-51.

Kroger, (2013a. Kroger website www.kroger.com

Kroger. (2013b). "2013 Sustainability Report."

Krull, C. F., Prescott, J. M. & Crum, C. W. (1998). "Seed Marketing and Distribution," *Maize Seed Industries in Developing Countries.* M.L. Morris (ed.). Boulder, CO: Lynne Rienner Pub./ CIMMYT.

Li, Q., Curtis, K. R., McCluskey, J. J. & Wahl, T. I. (2002). "Consumer Attitudes Toward Genetically Modified Foods in Beijing, China," *AgBioForum*, *5*(4), 145-162.

Lin, W., Somwaru, A., Tuan, F., Huang, J. & Bai, J. (2006). "Consumers' Willingness to Pay for Biotech Foods in China: A Contingent Valuation Approach," *AgBioForum*, *9*(3), 166-179.

List, J. A. & Gallet, C. A. (2001). "What Experimental Protocols Influence Disparities Between Actual and Hypothetical Stated Values?" *Environmental and Resource Economics*, *20*(3), 241-54.

Livingston, M. J., Carlson, G. A. & Fackler, P. L. (2004). "Managing Resistance Evolution in Two Pests to Two Toxins with Refugia," *American Journal of Agricultural Economics*, *86*(1), 1-13.

Livingston, M. J., Storer, N. P., Van Duyn, J. W. & Kennedy, G. G. (2007). "Do Refuge Requirements for Biotechnology Crops Promote Economic Efficiency? Some Evidence for Bt Cotton," *Journal of Agricultural and Applied Economics*, *39*(1), 171-185.

Loureiro, M. L. & Hine, S. (2002). "Discovering Niche Markets: A Comparison of Consumer Willingness to Pay for Local (Colorado Grown), Organic, & GMO-Free Products," *Journal of Agricultural and Applied Economics*, *34*(3), 477-87.

Lu, Y., Wu, K., Jiang, Y., Guo, Y. & Desneux, N. (2012). "Widespread Adoption of Bt Cotton and Insecticide Decrease Promotes Biocontrol Services," *Nature*, (July 19)(487)(7407), 362-5.

Lusk, J. L., Jamal, M., Kurlander, L., Roucan, M. & Taulman, L. (2005). "A Meta Analysis of Genetically Modified Food Valuation Studies." *Journal of Agricultural and Resource Economics*, *30*(April), 28-44.

Lusk, J. L. (2003a). "Effects of Cheap Talk on Willingness to Pay for Golden Rice,". *American Journal of Agricultural Economics*, *85*(4)(November), 840-856.

Lusk, J. L., Roosen, J. & John Fox. (2003). "Demand for Beef with Growth Hormones and Fed Genetic Corn," *American Journal of Agricultural Economics*, *85*(1)(Feb.), 16-29.

Maddala, G. S. (1983). *Limited Dependent and Qualitative Variables in Econometrics*. Cambridge, UK: Cambridge University Press.

Mather, D.W., Knight, J. G., Insch, A., Holdsworth, D. K., Ermen, D. F. & Breitbarth, T. (2011). "Social Stigma and Consumer Benefits: Trade-Offs in Adoption of Genetically Modified Foods," *Science Communication*.

Marra, M. C. & Piggott, N. E. (2006). "The Value of Non-Pecuniary Characteristics of Crop Biotechnologies: A New Look at the Evidence," *Regulating Agricultural Biotechnology: Economics and Policy*. In R. Just, J. Alston, & D. Zilberman (Eeds.). New York: SpringerVerlag Pub.

Marra, M., Carlson, G. & Hubbell, B. (1998). "Economic Impacts of the First Crop Biotechnologies." North Carolina State University. http://www.ag. econ.ncsu.edu/faculty/marra/ firstcrop/imp001.gif

Marra, M. C., Pardey, P. G. & Alston, J. M. (2002). "The payoffs to transgenic field crops: An assessment of the evidence,." *AgBioForum, 5*(2), 43-50.

Marra, M. C., Pardey, P. G. & Alston, J. M. (2002a). *The Payoffs to Agricultural Biotechnology: An Assessment of the Evidence*. EPTD Discussion Paper No 87. International Food Policy Research Institute. Washington, DC. Jan.

Marra, M., Piggott, N. E. & Carlson, G. A. (2004(. *The Net Benefits, Including Convenience of Roundup Ready Soybeans: Results from a National Survey*. NSF Center for IPM.

Martinez-Poveda, A., Brugarolas Molla-Bauza, M., del Campo Gomis, F. J. & MartinezCarrasco Martinez, L. (2009). "Consumer-perceived Risk Model for the Introduction of Genetically Modified Food in Spain," *Food Policy*, Vol. *34*(6), 519-528.

McBride, W. D. & El-Osta, H. (2002). "Impacts of the Adoption of Genetically Engineered Crops on *Journal of Agricultural and Applied Economics*, *34*(1), 175-191.

Mesnage, R., Bernay, B. & Séralini, G.-E. (2012). "Ethoxylated adjuvants of glyphosate-based herbicides are active principles of human cell toxicity," *Toxicology*.

Miranowski, J. A. & Carlson. G. A. (1986). "Economic Issues in Public and Private Approaches to Preserving Pest Susceptibility," *Pesticide Resistance: Strategies and Tactics for Management*, National Academies Press, Washington, DC.

Monsanto. (2012). *Research and Development Pipeline* http://www.monsanto.com/products/pages/research-development-pipeline.aspx

Moon, W., Balasubramanian, S. K. & Rimal, A. (2007). "Willingness to Pay (WTP) a Premium for Non-GM Foods versus Willingness to Accept (WTA) a Discount for GM Foods," *Journal of Agricultural and Resource Economics*, *32*(2), (Aug.), 363-82.

Moon, Wan Ki. & Balasubramanian, S. K. (2001). "Public Perceptions And Willingness-To-Pay A Premium For Non-GM Foods In The US And UK," *AgBioForum*, *4*(3&4), 221-231.

Mueller, T. C., Mitchell, P. D., Young, B. G. & Culpepper, A. S. (2005). "Proactive Versus Reactive Management of Glyphosate-Resistant or -Tolerant Weeds," *Weed Technology*, *19*(4), 924-933.

Mungai, N., Motavalli, P., Nelson, K. & Kremer, R. (2005). "Differences in yields, residue composition and N mineralization dynamics of Bt and non-Bt maize," *Nutrient Cycling in Agroecosystems*, (*73*), 101-109.

Murphy, J., Allen, P., Stevens, T. & Weatherhead, D. (2005). "A Meta-Analysis of Hypothetical Bias in Stated Preference Valuation," *Environmental and Resource Economics*, *30*(3), 313-25.

National Research Council (NRC). (2002. *Environmental Effects of Transgenic Plants: The Scope and Adequacy of Regulation.* Board on Agriculture and Natural Resources, Committee on Environmental Impacts Associated with Commercialization of Transgenic Plants. Washington, DC: The National Academies Press.

National Research Council (NRC). (2010. *The Impact of Genetically Engineered Crops on Farm Sustainability in the United States*, Washington, DC: National Academies Press.

National Research Council (NRC). (2012). *National Summit on Strategies to Manage HerbicideResistant Weeds.* Proceedings of a Workshop Organized by the Planning Committee for a National Summit on Strategies To Manage Herbicide-Resistant Weeds. Board of Agriculture and Natural Resources; Division of Earth and Life Studies. http://www.nap.edu/ catalog. php?record_id=13518

Nolan, E. & Santos, P. (2012). "The Contribution of Genetic Modification to Changes in Corn Yield in the United States," *American Journal of Agricultural Economic*, *94*(5), 1171-1188.

Norsworthy, J. K., Ward, S., Shaw, D., Llewellyn, R., Nichols, R., Webster, T., Bradley, K., Frisvold, G., Powles, S., Burgos, N., Witt, W. & Barrett, M. (2012). "Reducing the Risks of Herbicide Resistance: Best

Management Practices and Recommendations," *Weed Science*, *60* (Special Issue), 31-62.

Noussair, C., Robin, S. & Ruffieux, B. (2004). "Do Consumers Really Refuse to Buy Genetically Modified Food? " *Economic Journal*, *114*(492), (Jan.), 102-20.

Onyango, B., Nayga, R. M. Jr. & Schilling, B. (2004). "Role of Product Benefits and Potential Risks in Consumer Acceptance of Genetically Modified Foods," *AgBioForum*, *7*(4), 202-2011.

Osteen, C. & Fernandez-Cornejo, J. (2013). "Economic and Policy Issues of U.S. Agricultural Pesticide Use Trends," *Pest Management Science*, *69*(9), (Sept.), 1001-1025.

Parker, R. G., York, A. C. & Jordan, D. L. (2005). "Comparison of Glyphosate Products in Glyphosate-Resistant Cotton and Corn," *Weed Technology*, (19), 796-802.

Parker, R. G., York. A. C. & Jordan, D. L. (2006). "Weed control in glyphosate-resistant corn as affected by preemergence herbicide and timing of postemergence herbicide application," *Weed Technology*, *20*, 564-570.

Pew Initiative on Food and Biotechnology. (2001). *Harvest on the Horizon: Future Uses of Agricultural Biotechnology*. http://pewagbiotech.org/research/harvest/harvest.pdf. *Sept.*

Piggott, N. E. & Marra, M. C. (2007). "The net gain to cotton farmers of a natural refuge plan for Bollgard II® Cotton," *AgBioForum*, *10*(1), 1-10.

Pilcher, C. D., Rice, M. E., Higgins, R. A., Steffey, K. L., Hellmich, R. L., Witkowski, J., Calvin, D., Ostlie, K. R. & Gray, M. (2002). "Biotechnology and the European Corn Borer: Measuring Historical Farmer Perceptions and Adoption of Transgenic Bt Corn as a Pest Management Strategy." *Journal of Economic Entomology*, *95*(5), 878-892.

Pioneer Hi-Bred (A DuPont Business). (2012). *Research and Technology Pipeline.*

Publix. (2013). http://www.publix.com/about/qa/QAProducts.do#Q5

Reddy, K. N. & Whiting, K. (2000)."Weed control and economic comparisons of glyphosate-resistant, sulfonylurea-tolerant, & conventional soybean (Glycine max) systems," *Weed Technology*, (14), 204-211.

Regev, U., Shalit, H. & Gutierrez. A. P. (1983). "On the Optimal Allocation of Pesticides with Increasing Resistance: The Case of the Alfalfa Weevil," *Journal of Environmental Economics and Management*, *10*, 86-100.

ReJesus, R. M., Greene, J. K., Hamming, M. D. & Curtis, C. E. (1997). "Economic Analysis of Insect Management Strategies for Transgenic Bt Cotton Production in South Carolina," *Proceedings of the Beltwide Cotton Conferences*. Cotton Economics and Marketing Conference.

Rice, M. E. & Pilcher, C. D. (1998). "Potential Benefits and Limitations of Transgenic Bt Corn for Management of the European Corn Borer (*Lepidoptera: Crambidae*)," *American Entomologist*, (*44*), 75-78.

Roberts, R. K., Pendergrass, R. & Hayes, R. M. (1998). "Farm-Level Economic Analysis of Roundup Ready® Soybeans." Paper presented at the Southern Agricultural Economics Association Meeting, Little Rock, AR, Feb. 1-4.

Rousu, M. C. & Lusk, J. L. (2009). "Valuing Information on GM Foods in a WTA Market: What Information is Most Valuable? " *AgBioForum*, (*12*), 226-231.

Rousu, M., Huffman, W. E., Shogren, J. F. & Tegene, A. (2004). "Are United States Consumers Tolerant of Genetically Modified Foods? " *Review of Agricultural Economics*, *26*(Feb.), 19-31.

Runge, C. Ford, & Ryan, B. (2004). *The Global Diffusion of Plant Biotechnology: International Adoption and Research in 2004*. University of Minnesota.

Safeway/Vons. (2013). "Statement on Genetically Modified Organisms, October 4, 2013," http://www.vons.com/CMS/assets/media/AboutThisSite/StatementonGeneticallyModifiedOrganisms_Oct2013.pdf

Sainsbury. (2013. http://www.j-sainsbury.co.uk/extras/faqs/responsibility/sourcing/what-is-sainsburys-policy-on-gm/

Samsel, A. & Seneff, S. (2013). Glyphosate's Suppression of Cytochrome P450 Enzymes and Amino Acid Biosynthesis by the Gut Microbiome: Pathways to Modern Diseases. *Entropy*, *15*(4), 1416-1463.

Séralini, G. E., Clair, E., Mesnage, R., Gress, S., Defarge, N., Malatesta, M., Hennequin, D. & Spiroux de Vendômois, J. (2012). "Long term toxicity of a Roundup herbicide and a Rounduptolerant genetically modified maize," *Food and Chemical Toxicology*, *50*(11), 4221-4231.

Shaw, D. R., Owen, M. D. K., Dixon, P. M., Weller, S. C., Young, B. G., Wilson, R. G. & Jordan. D. L. 2011. "Benchmark Study on Glyphosate-Resistant Cropping Systems in the United States. Part 1: Introduction to 2006-2008," *Pest Management Science*, (*67*), 741-746.

Shi, G., Chavas, J.-P. & Stiegert, K. W. (2008). *An Analysis of Bundle Pricing: The Case of the Corn Seed Market*. FSWP2008-01. University of Wisconsin-Madison, Madison, WI. Nov.

Shi, G., Chavas, J.-P. & Stiegert, K. W. (2009). "Pricing of Herbicide-Tolerant Seeds: A Market Structure Approach," *AgBioForum*, *12*(3&4), 326-333.

Shi, G., Chavas, J.-P. & Stiegert, K. (2010). "An Analysis of the Pricing of Traits in the U.S. Corn Seed Market," *American Journal of Agricultural Economics*, *92*(5), 1324-1338.

Shi, G., Chavas, J.-P., Lauer, J. & Nolan, E. (2013)."An Analysis of Selectivity in the Productivity Evaluation of Biotechnology: An Application to Corn," *American Journal of Agricultural Economics*, *95*(3), 739-754.

Stiegert, K. W., Shi, G. & Chavas, J.-P. (2010). "Innovation, Integration and the Biotechnology Revolution in U.S. Seed Markets," *Choices*, *25*(2).

Stark, C. R. Jr. (1997). "Economics of Transgenic Cotton: Some Indications Based on Georgia Producers," *Beltwide Cotton Conference Proceedings*, Cotton Economics and Marketing Conference.

Southern Hemisphere Association of Fresh Fruit Exporters (SHAFE). (2011). *Guide to Environmental Requirements, Social, and Good Agricultural Practice (GAP) for Supermarket Chains*. For Chilean Exporters Association, June.

Tabashnik, B. E., Gassman, A. J., Crowder, D. W. & Carrière, Y. (2008). "Insect Resistance to Bt Crops: Evidence Versus Theory," *Nature Biotechnology*, *26*(2), 199-202.

Tabashnik, B. E., Brévault T. & Carrière, Y. (2013). "Insect Resistance to Bt crops: Lessons from the First Billion Acres. *Nature Biotechnology*, *31*(6), 510-521.

Tegene, A., Huffman, W., Rousu, M. & Shogren, J. (2003). *The Effects of Information on Consumer Demand for Biotech Foods: Evidence from Experimental Auctions*. U.S. Department of Agriculture, Economic Research Service, TB-1903, April.

Tesco. (2011). http://www.tesco.com/greenerliving/greener_tesco/what_tesco_is_doing/sustainable_living.page

Tesco. (2013) "An Update on Poultry Feed," *Food News*. April. http://tescofoodnews.com/news/an-update-on-our-poultry-feed/

Toevs, E .A., Guenthner, J. F., Johnson, A. J., McIntosh, C. S. & Thornton, M. K. (2011). "An Industry Perspective of All-Native and Transgenic Potatoes," *AgBioForum*, *14*(1), 14-19.

U.S. Department of Agriculture (USDA), Agricultural Research Service (ARS). (2012). *ARS Timeline:144 Years of Research*. http://www.ars.usda.gov/is/timeline/1980chron.htm?pf=1

U.S. Department of Agriculture (USDA), Animal and Plant Health Inspection Service (APHIS). 2004. *Changes in Regulation of the Products of Biotechnology by APHIS' Biotechnology Regulatory Services.* http://www.aphis.usda.gov/brs/eis/index.html. Jan.

USDA Advisory Committee on Biotechnology and 21st Century Agriculture (AC21). (2012). *Enhancing Coexistence: A Report of the AC21 to the Secretary of Agriculture November 19,* http://www.usda.gov/documents/ac21_report-enhancing-coexistence.pdf

U.S. Department of Agriculture (USDA), Animal and Plant Health Inspection Service (APHIS). (2011). *User Guide: Notification.* Biotechnology Regulatory Services. March 29 http://www.aphis.usda.gov/biotechnology/downloads/notificatiguidanceon_guidance_0311.pdf

U.S. Department of Agriculture (USDA), Animal and Plant Health Inspection Service (APHIS). (2012). *Biotechnology: Biotechnology Regulatory* Services. http://www.aphis.usda.gov/biotechnology/brs_main.shtml

U.S. Department of Agriculture (USDA), Animal and Plant Health Inspection Service (APHIS). (2013). *Biotechnology: BRS Federal Register Notices by Topic.* http://www.aphis.usda.gov/biotechnology/fr_notices_topic.shtml

U.S. Department of Agriculture (USDA), Economic Research Service (ERS) in collaboration with the USDA Natural Resources Conservation Service (NRCS). (1998). *Economic and Environmental Benefits and Costs of Conservation Tillage.* Feb.

U.S. Department of Agriculture (USDA), Economic Research Service (ERS). (2013). Organic Production Data. http://www.ers.usda.gov/data-products/organic-production.aspx#.UrSN_vv9zjM

U.S. Department of Agriculture (USDA), Economic Research Service (ERS). (2013). *Adoption of Genetically Engineered Crops in the U.S.* Data Product, http://www.ers.usda.gov/Data/ BiotechCrops/ July.

U.S. Department of Agriculture (USDA), National Agricultural Statistics Service (NASS). Various years. *Agricultural Prices.*

U.S. Environmental Protection Agency (USEPA). (2003). *EPA's Regulation of Biotechnology for Use in Pest Management.* http://www.epa.gov/pesticides/biopesticides/reg_of_biotech/ eparegofbiotech.htm. June.

Vencill, W. K. (1996). "Weed Management Systems Utilizing Herbicide-resistant Cotton," *Proceedings of the Beltwide Cotton Conferences.* Cotton Weed Science Research Conference.

Volinskiy, D., Adamowicz, W. L., Veeman, M. & Srivastava, L. (2009). "Does Choice Context Affect the Results from Incentive-Compatible

Experiments? The Case of Non-GM and Country-of-Origin Premia in Canola Oil," *Canadian Journal of Agricultural Economics*, *57*(2), (June), 205-21.

Walmart. (2013a). www.walmart.com

Walmart. (2013b). "2013 Global Responsibility Report." http://corporate.walmart.com/global-responsibility/environment-sustainability/global-responsibility-report

West, G., Gendron, C., Larue, B., & Lambert, R. (2002). "Consumers' Valuation of Functional Properties of Foods: Results from a Canada-Wide Survey," *Canadian Journal of Agricultural Economics*, *50*(4) (Dec.), 541-58.

World Health Organization (WHO). (1994). *Glyphosate*. International Programme of Chemical Safety (IPCS), Geneva. Environmental Health Criteria No. 159.

Zaid, A., Hughes, H. G., Porceddu, E. & Nicholas, F. (1999). *Glossary of Biotechnology and Genetic Engineering*. FAO Research and Technology Paper. Food and Agriculture Organization of the United Nations. Rome.

End Notes

[1] Genetic engineering is a technique used to alter genetic material (genes) of living cells. A gene is a segment of DNA that expresses a particular trait. It is a unit of heredity transmitted from generation to generation during reproduction (Zaid et al., 1999). DNA constitutes the genetic material of most known organisms.

[2] Plant biotechnology in general and genetic engineering in particular have significantly reduced the time needed to develop improved plant varieties, increasing the range and precision of characteristics incorporated into these new varieties (Fernandez-Cornejo, 2004). By allowing scientists to target single plant traits through genetic recombination techniques, plant biotechnology decreases the number of residual unwanted characteristics that often result from traditional plant breeding crosses, enabling breeders to develop desirable new varieties more rapidly.

[3] Several second-generation GE crops have been approved by APHIS: high-lysine corn, reduced-nicotine tobacco, higholeic acid soybean oil, stearidonic acid-producing soybeans, improved fatty acid-profile soybeans, altered-flower color roses (blue), oil profile-altered canola, and alpha amylase corn. Overall, nearly 20 percent of the approvals for deregulation (as of September 2013) are second-generation crops.

[4] Another indicator of R&D activity is the number of patents issued by the U.S. Patent and Trademark Office. More than 4,200 new agricultural biotech patents were issued between 1996 and 2000 (King and Heisey, 2003, 2004).

[5] A gene construct is the technical name used for a functional unit necessary for the transfer or the expression of a gene of interest (http://www.gmo-safety.eu/glossary/667.gene-construct.html). Typically, a construct comprises the gene or genes of interest, a marker gene (to facilitate detection inside the plant), and appropriate control sequences as a single package (FAO, 2001).

[6] Within each location there can be multiple sites or fields where the trial will be carried out (Information Systems for Biotechnology, 2013).

[7] A petition (as well as an approval) may include more than one trait or phenotype category. For example, a petition for corn may include one or more HT traits and one or more Bt traits.

[8] Pharmaceutical plant compounds produced are intended for pharmaceutical use and would need to be approved from at least one of the following agencies prior to commercialization: U.S. Food and Drug Administration (FDA) Center for Biologics Evaluation and Research (human biologics), FDA Center for Drug Evaluation and Research (human drugs), FDA Center for Veterinary Medicine (animal drugs), and USDA Center for Veterinary Biologics (animal biologics). None of the plants currently under permit produce pharmacologically active compounds.

[9] Over 90 percent of U.S. acreage devoted to major crops has been treated with herbicides in recent decades (Osteen and Fernandez-Cornejo, 2012).

[10] Some other GE crops were only on the market for a limited amount of time. Bt potato varieties were introduced in 1996, but withdrawn from the market after the 2001 season. FlavrSavr tomatoes, which were genetically engineered to remain on the vine longer and ripen to full flavor after harvest, were introduced in 1994, but withdrawn from the market after several years.

[11] Drought-tolerant corn was approved for commercial use in 2011 (*Federal Register*, 2011; Monsanto, 2012) and commercially introduced in 2012.

[12] USDA's National Agricultural Statistics Service (NASS) and Economic Research Service (ERS) sponsor the Agricultural Resource Management Survey (ARMS). This survey includes a crop-specific survey of production practices (called ARMS Phase 2) for selected major corps each once every 5 years on a rotating basis. This survey was conducted in 2006 for soybeans, 2007 for cotton, and 2010 for corn.

[13] Potential yield is defined as "the yield of an adapted cultivar when grown with the best management and without natural hazards such as hail, frost, or lodging, and without water, nutrient, or biotic stress limitations (water stress being eliminated by full irrigation or ample rainfall)" (Fischer and Edmeades, 2010). Farm level (actual or effective) yield is equal to potential yield minus the yield lost to pests or to other stresses.

[14] Since Bt and HT traits protect yield rather than increase potential yield, it is possible that in some cases the Bt and HT traits are not introduced in the highest yielding germplasm. Over time, this so-called "yield drag" usually disappears (NRC, 2010, Ch 3). On the other hand, Shi et al. (2013) show that the opposite situation may arise if GE genes are added more frequently to "high quality" germplasm. They call this situation genetic selectivity bias.

[15] In this report, net returns are defined as per-acre revenues minus per-acre variable costs. Revenues per acre are equal to crop yields times crop price. Per-acre variable input costs include pesticide, seed and labor costs. Seed costs paid by adopters of GE varieties include a technology fee. This measure of net returns is used because most of the financial impacts of adopting GE crops result from changes in crop yields, chemical costs, and increased seed costs. This measure is estimated using field-level data and captures the greatest influence that GE crop adoption would have on farm financial performance as it also filters out the impact of other farm activities—such as livestock production (Fernandez-Cornejo and McBride, 2002). The econometric estimation involves estimating a restricted profit function (Fernandez-Cornejo and Wechsler, 2012) together with the associated supply function and input demand functions (hired labor is also included and wages are used as the *numeraire*).

[16] The difference in means of corn yields between adopters and nonadopters is statistically significant for 2005 and 2010 using either the delete-a-group jackknife procedure (Kott, 1998) or the standard statistical test.

[17] The panel members who wrote the NRC report were Y. Carriere, W. Cox, D. Ervin, J. Fernandez-Cornejo, R. Jussaume Jr., M. Marra, M. Owen, P. Raven, L. Wolfenbarger and D. Zilberman.

[18] Marra et al. (2002a) provides an extensive discussion of the various types of biases that can arise when comparing means not only in farm (and field) surveys but in experimental settings as well (see box 3 for a discussion of the bias that may be caused by the halo effect).

[19] Self-selection is a type of endogeneity (Maddala, 1983; Greene, 1997). Endogeneity arises when there is a correlation between the explanatory variable and the model's residuals. If endogeneity is not accounted for (for instance, through the use of instrumental variable techniques), the results of the analysis will be biased. A common approach used to control for self-selection is sometimes called an instrumental variables approach. The model includes two stages. The first stage, which is referred to as the *adoption decision model*, is used to estimate the predicted values of the probability of adoption using a probit model. The second stage, or *impact model*, uses the predictions estimated in the first stage to estimate the impact of adopting Bt seeds on yields, seed demand, insecticide demand, and net returns.

[20] Monsanto discontinued breeding and field level research on its GE Roundup Ready wheat in 2004.

[21] Because pest pressure varies from one region to another, the economic benefits of Bt corn and consequently the rates of adoption vary regionally (figure 6). Additionally, farmers must decide whether or not to use Bt corn before they know the severity of pest infestations, corn prices, or the price of insecticides. "Overadoption" may result from incorrect predictions (Fernandez-Cornejo and McBride, 2002). Alternately, farmers may be willing to adopt Bt seeds in order to reduce the risks associated with infestation levels that are higher than expected.

[22] Elasticity measures the responsiveness of a variable (e.g., net returns) to a change in another (e.g., adoption rate). It is unit free and is expressed in percentage terms.

[23] Given the high rates of adoption of HT soybeans (more than 90 percent in recent years), econometric studies using recent data are problematic because of the small size of the sample of nonadopters and the likelihood that there may be other factors influencing the decision not to adopt (e.g., organic farming) of that small group. This may lead to a stronger selection bias compared to studies using data from earlier years.

[24] Adoption of conservation tillage by HT adopters may have also confounded these comparisons.

[25] However, recent publications have raised questions regarding the toxicity of glyphosate. Seralini et al. (2012) claim that GE corn and low levels of glyphosate herbicide formulations at concentrations well below officially-set safe limits induce severe adverse health effects, such as tumors, in rats. But a review of the study by the European Food Safety Authority (EFSA, 2012) concluded the Seralini et al. study as reported in the publication "is inadequately designed, analyzed and reported" and is "of insufficient scientific quality for safety assessments. As a result, the EFSA states that "conclusions cannot be drawn on the difference in tumour incidence between treatment groups on the basis of the design, the analysis and the results as reported." In a separate study, Mesnage et al. (2012) find that while toxicity of glyphosate has been safety tested on mammals, another ingredient used in commercial formulations used as adjuvant is toxic. More recently, Samsel and Sanoff (2013) claim that "glyphosate enhances the damaging effects of other foodborne chemical residues and environmental toxins."

[26] HT crops also simplify the management of pesticide application (Carpenter and Gianessi, 1999). There is a greater window over which the herbicides can be applied (glyphosate can be effective on older plants). This makes it much easier to manage weather-related delays to the herbicide application schedule. Use of glyphosate also may reduce the need for aerial applications that are sometimes needed when it is too wet to enter the field.

[27] No-till systems are often considered the most effective of all conservation tillage systems. They leave 100 percent of crop residues on the soil surface and the soil is undisturbed from harvest to planting, resulting in the highest percentage of surface being covered by crop residues, minimizing soil loss and water runoff (Janssen and Hill, 1994).

[28] There is some indication of emergence of Bt-resistant corn rootworm in some parts of the Corn Belt http://www. bloomberg.com/news/2012-09-04/-mounting-evidence-of-bug-resistant-corn-seen-by-epa.html. http://bulletin.ipm. illinois.edu/article.php?id=1704. There is also anecdotal evidence that resistance is a contributing factor to increasing corn insecticide sales in 2012 and 2013 (I. Berry; *WSJ*, May 21, 2013). Tabashnik et al. (2013) recently analyzed 77 studies carried out in 5 continents from 1996 to 2012. They find that "although most pest populations remained susceptible, reduced efficacy of Bt crops caused by field-evolved resistance has been reported now for some populations of 5 of 13 major pest species examined, compared with resistant populations of only one pest species in 2005." They conclude that "the increase in documented cases of resistance likely reflects increases in the area planted to Bt crops, the cumulative duration of pest exposure to Bt crops, the number of pest populations exposed and improved monitoring efforts." They also conclude that while "regulations in the United States and elsewhere mandate refuges of non-Bt host plants for some Bt crops, farmer compliance is not uniformly high and the required refuge percentages may not always be large enough to achieve the desired delays in evolution of resistance. Both in theory and practice, using Bt crops in combination with other tactics as part of integrated pest management may be especially effective for delaying pest resistance."

[29] A dominant trait will be expressed in progeny if at least one of the parents has the gene for that trait. A recessive trait will be inherited if both parents have the gene for that trait. (Hedrick, 2000).

[30] According to Andow et al. (2008), "a structured refuge is one that is planted near Bt cotton deliberately and an unstructured refuge relies on the other crops already grown as part of the local cropping system and where Bt is not used."

[31] Impending expiration of glyphosate patent protection in 2000 and the availability of generic glyphosate herbicides have led to a decrease in its price since 1998.

[32] Dauer et al. (2009) provide empirical evidence that horseweeds, one of the more important glyphosate-resistant weed species, can disperse between farms.

[33] BMPs include applying multiple herbicides with different modes of action at the recommended rates and developmental stages for the target weeds in the field, increasing the intensity of tillage to reduce the fraction of seeds that germinate, planting weed-free crop seed, scouting fields routinely, cleaning equipment to reduce the rate of introducing weeds to other fields, and preventing weed introductions by maintaining field borders (Norsworthy et al., 2012).

[34] Glyphosate resistance in weed species and biotypes in the United States is also due to glyphosate use in tree orchards, on roadsides, and on non-HT crops, where it is used before crops are planted and after they are harvested.

[35] Many farmers incorrectly assume there is no need to adopt BMPs because new herbicides will be available in the future (Norsworthy et al. 2012). In addition, the benefits of using BMPs occur in the future and are uncertain, as opposed to the certain increase in production costs.

[36] More information about this program is available at https://www.roundupreadyplus.com/Pages/Home.aspx.

[37] Non-GE foods can be more expensive if they cost more to produce, or if marketing streams must be kept separate.

[38] Consumers who asked about the GE fruit were told that it contained genes that caused it to produce a natural insecticide. GE fruit is not commercially available in these countries, and consumers who expressed surprise about this were told the fruit may have come from an experimental orchard (Mather et al., 2011).

[39] The supermarket chain Whole Foods has announced a labeling policy that will be implemented by 2018 to indicate if their products contain GE ingredients (http://media.wholefoodsmarket.com/news/whole-foods-market-commits-to-fullgmo-transparency). ERS researchers discuss the economic issues related to food labeling, including GE foods (Golan et al., 2001).

[40] Tesco, ASDA, Sainsbury's, and William Morrison.

[41] WalMart, Kroger, Safeway, and Publix.

[42] Via labels or promotional material. The current website was sometimes consulted if the claim was ambiguous.

[43] http://www.aphis.usda.gov/newsroom/2013/09/ac21.shtml)

In: Genetically Engineered Crops in America ISBN: 978-1-63321-225-1
Editor: Fredrick G. Lawrence

Chapter 2

THE FIRST DECADE OF GENETICALLY ENGINEERED CROPS IN THE UNITED STATES*

Jorge Fernandez-Cornejo and Margriet Caswell
U.S. Department of Agriculture (USDA)

ABSTRACT

Ten years after the first generation of genetically engineered (GE) varieties became commercially available, adoption of these varieties by U.S. farmers is widespread for major crops. Driven by farmers' expectations of higher yields, savings in management time, and lower pesticide costs, the adoption of corn, soybean, and cotton GE varieties has increased rapidly. Despite the benefits, however, environmental and consumer concerns may have limited acceptance of GE crops, particularly in Europe. This chapter focuses on GE crops and their adoption in the United States over the past 10 years. It examines the three major stakeholders of agricultural biotechnology and finds that (1) the pace of R&D activity by producers of GE seed (the seed firms and technology providers) has been rapid, (2) farmers have adopted some GE varieties widely and at a rapid rate and benefited from such adoption, and (3) the level of consumer concerns about foods that contain GE

* This is an edited, reformatted and augmented version of a report, EIB-11, issued by the U.S. Department of Agriculture, Economic Research Service, April 2006.

ingredients varies by country, with European consumers being most concerned.

Keywords: Genetically engineered crops, agricultural biotechnology, seed industry, research and development, adoption, crop yields, pesticide use, corn, soybeans, cotton

ACKNOWLEDGMENTS

The authors wish to thank William Lin, Paul Heisey, Keith Wiebe, Marca Weinberg, Utpal Vasavada, and Mary Bohman of ERS for the helpful comments provided on earlier drafts of this report. We also thank Andrew Rude from the Foreign Agricultural Service, Neil E. Hoffman from the Animal and Plant Health Inspection Service, John W. Radin from the Agricultural Research Service, Daniel Jones from the Cooperative State Research, Education, and Extension Service, and Matthew C. Rousu from Susquehanna University. We also thank Lou King for editorial assistance and Anne Pearl for graphics and layout.

SUMMARY

Over the past decade, developments in modern biotechnology have expanded the scope of biological innovations by providing new tools for increasing crop yields and agricultural productivity. The role that biotechnology will play in agriculture in the United States and globally will depend on a number of factors and uncertainties. What seems certain, however, is that the ultimate contribution of agricultural biotechnology will depend on our ability to identify and measure its potential benefits and risks.

What Is the Issue?

Ten years after the first generation of genetically engineered (GE) varieties of major crops became commercially available, adoption of these varieties by U.S. farmers has become widespread. United States consumers eat many products derived from these crops—including some cornmeal, oils, sugars, and other food products—largely unaware of their GE content. Despite the rapid increase in the adoption of GE corn, soybean, and cotton varieties by U.S. farmers, questions remain regarding the impact of agricultural biotechnology. These issues range from the economic and environmental impacts to consumer acceptance.

What Did the Study Find?

This study examined the three major stakeholders in agricultural biotechnology: **seed suppliers and technology providers, farmers, and consumers.**

Seed suppliers/technology providers. Strengthening of intellectual property rights protection in the 1970s and 1980s increased returns to research and offered greater incentives for private companies to invest in seed development and crop biotechnology. Since 1987, seed producers have submitted nearly 11,600 applications to USDA's Animal and Plant Health Inspection Service for field testing of GE varieties. More than 10,700 (92 percent) have been approved. Approvals peaked in 2002 with 1,190. Most approved applications involved major crops, with nearly 5,000 for corn alone, followed by soybeans, potatoes, and cotton. More than 6,600 of the approved applications included GE varieties with herbicide tolerance or insect resistance. Significant numbers of applications were approved for varieties with improved product quality, viral resistance, and enhanced agronomic properties such as drought and fungal resistance.

Farmers. Adoption of GE soybeans, corn, and cotton by U.S. farmers has increased most years since these varieties became commercially available in 1996. By 2005, herbicide-tolerant soybeans accounted for 87 percent of total U.S. soybean acreage, while herbicide-tolerant cotton accounted for about 60 percent of total cotton acreage. Adoption of insect-resistant crops is concentrated in areas with high levels of pest infestation and varies across States. Insect-resistant cotton was planted on 52 percent of cotton acreage in 2005—ranging from 13 percent in California to 85 percent in Louisiana. Insect-resistant corn accounted for 35 percent of the total acreage in 2005, following the introduction of a new variety to control the corn rootworm.

The economic impact of GE crops on producers varies by crop and technology. Herbicide-tolerant cotton and corn were associated with increased returns, as were insect-resistant cotton and corn when pest infestations were more prevalent. Despite the rapid adoption of herbicide-tolerant soybeans, there was little impact on net farm returns in 1997 and 1998. However, the adoption of herbicide-tolerant soybeans is associated with increased off- farm household income, suggesting that farmers adopt this technology because the simplicity and flexibility of the technology permit them to save management time, allowing them to benefit from additional income from off-farm activities.

Genetically engineered crops also seem to have environmental benefits. Overall pesticide use is lower for adopters of GE crops, and the adoption of herbicide-tolerant soybeans may indirectly benefit the environment by encouraging the adoption of soil conservation practices.

Consumers. Most surveys and consumer studies indicate consumers have at least some concerns about foods containing GE ingredients, but these concerns have not had a large impact on the market for these foods in the United States. Despite the concerns of U.S. consumers, "GE-free" labels on foods are not widely used in the United States. Manufacturers have been active in creating a market for GE-free foods. Between 2000 and 2004, manufacturers introduced more than 3,500 products that had explicit non- GE labeling, most of them food products.

In the European Union and some other countries, however, consumer concerns have spurred a movement away from foods with GE ingredients. Despite the fact that some European consumers are willing to consume foods containing GE ingredients, very few of these foods are found on European grocery shelves.

How Was the Study Conducted?

This chapter examined the three major stakeholders of agricultural biotechnology: GE seed suppliers and technology providers (biotech firms), farmers, and consumers. To examine biotech and seed firms, we used information from the literature as well as from the database of USDA approvals of field testing for new GE varieties. To study seed users, we drew on ERS studies based on USDA farm surveys, and to review the consumer perspective, we summarized surveys of consumers' attitudes from the literature.

Introduction

Over the past decade, developments in modern biotechnology have expanded the scope of biological innovations by providing new tools for increasing crop yields and agricultural productivity. Agricultural biotechnology is a collection of scientific techniques, including genetic engineering, that are used to create, improve, or modify plants, animals, and microorganisms. Genetic engineering (GE) techniques allow a precise alteration of a plant's traits (facilitating the development of characteristics not possible through traditional plant breeding), and permit targeting of a single plant trait (decreasing the number of unintended characteristics that may occur with traditional breeding).[1]

The commercial success of GE crop varieties typically requires that biotechnology-derived trait enhancements be incorporated into successful cultivars (cultivated varieties with useful agronomic properties), the development of which requires significant knowledge of traditional plant breeding and the availability of genetic material (germplasm). This complementarity has been related to various institutional arrangements between seed and technology suppliers.

GE crops are often classified into one of three generations (Panos). Crops with enhanced input traits, such as herbicide tolerance, insect resistance, and tolerance to environmental stresses (like drought), represent the first generation. GE crops benefit farmers and may also offer environmental benefits. Second-generation crops include those with added-value output traits, such as nutrient enhancement for animal feed. Consumers will benefit directly from these products when they are available on the market. The third generation includes crops that produce pharmaceuticals or improve processing of bio-based fuels, and products beyond traditional food and fiber. At present, adoption of GE crops is generally limited to those with first-generation traits, which were tested on a large scale (field testing) in the 1980s to ensure that the desired traits will perform under production conditions. Second- and third-generation GE crops are in various stages of research and development.

Ten years after the first generation of GE varieties became commercially available, they have been widely adopted by U.S. farmers, driven by expectations of higher yields, savings in management time, and lower pesticide costs. Despite these benefits, environmental and consumer concerns may have limited acceptance of agricultural biotechnology, particularly in Europe. In the United States, foods containing GE ingredients currently available in the U.S. market do not require labels, since the U.S. Food and Drug Administration has determined that these foods are "substantially equivalent" to their non-GE counterparts (Shoemaker et al., 2003; FDA, 1992). Thus, U.S. consumers have been eating foods that contain GE ingredients (corn meal, oils, sugars) for the past 10 years while remaining largely unaware of their GE content.

Rapid Change and Pace of R&D Activity Characterize the Seed Industry and Technology Providers

The U.S. commercial seed market is the world's largest—with an estimated annual value of $5.7 billion per year in the late 1990s—followed by China at $3 billion and Japan at $2.5 billion (Fernandez-Cornejo, 2004). Moreover, the U.S. seed market is growing (in quantity and value), mainly because farmers have been increasing purchases of seed and reducing the planting of saved seed.

Growth in the seed market has been particularly rapid for major field crops—corn, soybeans, cotton, and wheat—that together constituted two-thirds of the seed market value in 1997.

The U.S. seed industry began a transformation in the 1930s, with the introduction of commercially viable hybrid seeds. These hybrids were higher yielding than nonhybrid varieties but degenerative, so farmers had to purchase new seed every year to maintain the high yields. Further changes were motivated by the strengthening of intellectual property rights (IPR) protection, mainly during the 1970s and 1980s, which increased returns to research and offered a greater incentive for private companies to invest in seed development. The two principal forms of legal protection are plant variety protection (PVP) certificates issued by the Plant Variety Protection Office of USDA and patents issued by the U.S. Patent and Trademark Office. Both grant private crop breeders exclusive rights to multiply and market their newly developed varieties.

However, patents provide more control since PVP certificates have a research exemption allowing others to use the new variety for research purposes. Agricultural biotechnology patents, mostly dealing with some aspect of plant breeding, have outpaced the general upward trend in patenting throughout the U.S. economy. During 1996-2000, 75 percent of over 4,200 new agricultural biotech patents went to private industry (King and Heisey).

Enhanced protection of intellectual property rights brought rapid increases in private research and development (R&D) investments and changes in market concentration in the U.S. seed industry. R&D expenditures on plant breeding for many major crops shifted from mainly public to mainly private. Private spending on crop variety R&D increased fourteenfold between 1960 and 1996 (adjusted for inflation), while public expenditures changed little (Fernandez-Cornejo, 2004).

As the amount of private capital devoted to R&D in the seed industry grew rapidly, the number of private firms engaged in plant breeding also grew, until peaking in the early 1990s. Subsequently, the seed industry consolidated, with fewer firms capable of sustaining the research investment needed to develop new seed varieties. Mergers and acquisitions created a seed industry structure dominated by large companies with primary investments in related sectors, such as pharmaceutical, petrochemical, and food (Fernandez-Cornejo, 2004).[2] In the early 1980s, developments in biological sciences created an additional incentive for private firms to increase their investment in R&D and seed production. As the first products of crop biotechnology were tested on a large scale in the 1980s, the seed industry's structure underwent additional transformation. Companies sought to achieve economies of scale to offset the high costs of biotechnology R&D through an extensive process of mergers, acquisitions, and joint ventures. Chemical and seed businesses combined to take advantage of strong demand complementarities between products (Just and Hueth, 1993). For example, the herbicide glyphosate and soybean seeds tolerant to glyphosate are sold by the same firm. As a consequence of the merger activity, the seed industry became more concentrated. By 1997, the share of U.S. seed sales (including GE and conventional varieties) controlled by the four largest firms providing seed of each crop reached 92 percent for cotton, 69 percent for corn, and 47 percent for soybeans (table 1).

From the Laboratory to the Field

A critical part of new variety development is field-testing to ensure that the desired traits will perform under production conditions. The release of new GE varieties of organisms into the environment is regulated through field release permits and monitored by USDA's Animal and Plant Health Inspection Service (APHIS) (see box, "Regulatory Oversight"). The number of field releases of plant varieties for testing purposes provides a useful indicator of R&D efforts on crop biotechnology.

Regulatory Oversight

Before commercial introduction, genetically engineered crops must conform to standards set by State and Federal statutes (USDA, 2005). Under the Coordinated Framework for the Regulation of Biotechnology, Federal oversight is shared by the U.S. Department of Agriculture (USDA), the U.S. Environmental Protection Agency (EPA), and the U.S. Food and Drug Administration (FDA).

USDA's Animal and Plant Health Inspection Service (APHIS) plays a central role in regulating field-testing of agricultural biotechnology products. Through either a notification or permit procedure, such products, which include genetically engineered plants, microorganisms, and invertebrates, are considered "regulated articles." APHIS determines whether to authorize the test, based on whether the release will pose a risk to agriculture or the environment. After years of field tests, an applicant may petition APHIS for a determination of nonregulated status in order to facilitate commercialization of the product. If, after extensive review, APHIS determines that the unconfined release does not pose a significant risk to agriculture or the environment, the organism is "deregulated." At this point, the organism is no longer considered a regulated article and can be moved and planted without APHIS authorization (USDA, 2004).

If a plant is engineered to produce a substance that "prevents, destroys, repels, or mitigates a pest," it is considered a pesticide and is subject to regulation by EPA (*Federal Register*, November 23, 1994). FDA regulates all food applications of crops, including those crops that are developed through the use of biotechnology, to ensure that foods derived from new plant varieties are safe to eat. A more complete description of the EPA and FDA regulations of GE products may be found in EPA (2003) and FDA (1992, 2005).

Though the current regulatory system is considered to be effective, USDA, EPA, and FDA continuously look forward and make necessary changes to address new trends and issues of the future. For example, USDA's APHIS has made updates in 1993 and 1997 and is currently considering the need for additional changes in the regulations (USDA, 2004).

The National Academy of Sciences also issued a report that made recommendations suggesting that regulation "could be improved further" by making the process more "transparent and rigorous" by enhanced scientific peer review, solicitation of public input, and "more explicit presentation of data, methods, analyses, and interpretations" (NRC, 2003).

Table 1. Estimated U.S. seed market shares for major field crops, 1997

Company	Corn	Soybean	Cotton
Pioneer Hi-Bred	42.0	19.0	
Monsanto[1]	14.0	19.0	11.0
Novartis	9.0	5.0	
Delta & Pine Land[2]			73.0
Dow Agrosciences/Mycogen	4.0	4.0	
California Planting Seed Distributors			6.0
All-Tex			2.0
Four largest total	69.0	47.0	92.0

[1] Monsanto acquired DeKalb in 1997 and Asgrow in 1998. [2] The merger proposed between Monsanto and Delta & Pine Land in 1998 was called off in December 1999. Source: Fernandez-Cornejo, 2004.

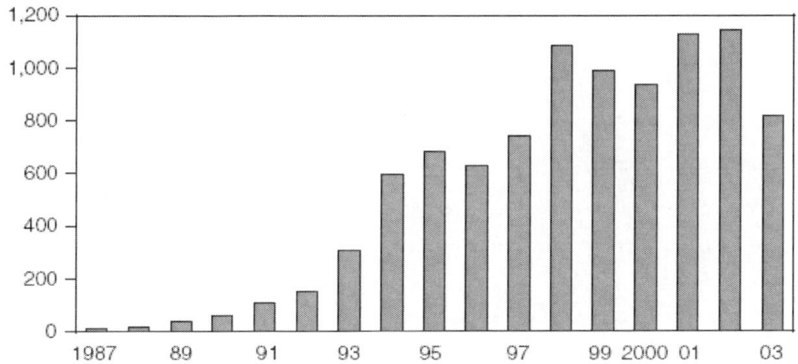

Source: Virginia Polytechnic Institute and State University, 2005.

Figure 1. Permits for release of GE varieties approved by APHIS.

By early April 2005, nearly 11,600 applications had been received by APHIS since 1987 and more than 10,700 (92 percent) had been approved (Virginia Polytechnic Institute and State University, 2005). Approvals peaked in 2002 with 1,190 (figure 1). Most applications approved for field testing involved major crops, particularly corn with nearly 5,000 applications approved, followed by soybeans, potatoes, cotton, tomatoes, and wheat (figure 2). Applications approved between 1987 and early April 2005 included GE varieties with herbicide tolerance (3,587), insect resistance (3,141), improved product quality (flavor, appearance, or nutrition) (2,314), virus resistance (1,239), and agronomic properties like drought resistance (1,043) and fungal resistance (647) (figure 3).

Table 2. Biotech crops currently available and in development in the United States

Crop	Input traits				Output traits	
	Herbicide tolerance	Insect resistance	Viral/fungal resistance	Agronomic properties[9]	Product quality[11]	Nutraceuticals; pharmaceuticals; industrial[13]
Corn	C	C[5]	D	D	D	D
Soybeans	C	D		D	D	
Cotton	C	C[6]		D	D	
Potatoes		W[7]	D	D	D	D
Wheat	C[2]		D			
Other field crops[1]	C[3] D[4]	D	D	D	D	D
Tomato, squash, melon			D	D	W[12] D	D
Other vegetables	D				D	
Papaya			C[8]			
Fruit trees			D		D	
Other trees				D[10]	D	
Flowers					D	

C = Currently available; D = In various stages of development and testing; W = Withdrawn from the market.

Sources: Virginia Polytechnic Institute and State University; USDA, APHIS; Colorado State University; Shoemaker et al.; Pew.

[1]Includes barley, canola, peanuts, tobacco, rice, alfalfa, etc.

[2]Monsanto discontinued breeding and field-level research on its GE Roundup Ready wheat in 2004.

[3]Canola.

[4]Barley, rice, sugar beets.

[5]Bt corn to control the corn borer commercially available since 1996; Bt corn for corn rootworm control commercially available since 2003.

[6]Bt cotton to control the tobacco budworm, the bollworm, and the pink bollworm, commercially available since 1996.

[7]Bt potatoes, containing built-in resistance to the Colorado potato beetle, were commercially introduced in 1996 and withdrawn in 1999.

[8]In the mid 1990s, researchers at Cornell University and at the University of Hawaii developed two virus-resistant varieties of GE papaya. First commercial plantings were made in 1998. The new varieties were proved successful in resisting a viral epidemic and were planted on more than 30 percent of Hawaii's papaya acreage in 1999.

[9]Resistance to cold, drought, frost, salinity; more efficient use of nitrogen; increased yield.

[10]Modified lignin content (for example, to reduce cost of paper making from trees).

[11]Includes delayed ripening; increased protein, carbohydrate, fatty acid, micronutrient, oil, and modified starch content; enhanced flavor and texture (fruits and vegetables); color (cotton, flowers); fiber properties (cotton); gluten content; natural decaffeination; and low phytase.

[12]Tomato genetically engineered to remain on the vine longer and ripen to full flavor after harvest; currently withdrawn from the market (Colorado State University, 2004).

[13]Includes increased vitamin, iron, beta-carotene content; antibodies, vaccines; specialty machine oils.

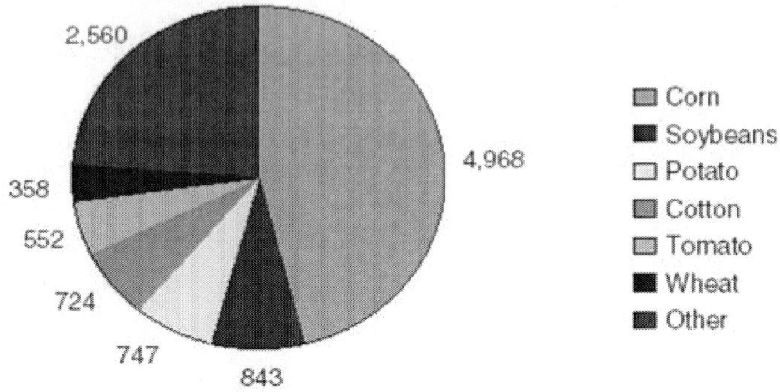

Source: Virginia Polytechnic Institute and State University, 2005.

Figure 2. Total number of permits approved by APHIS, by crop.

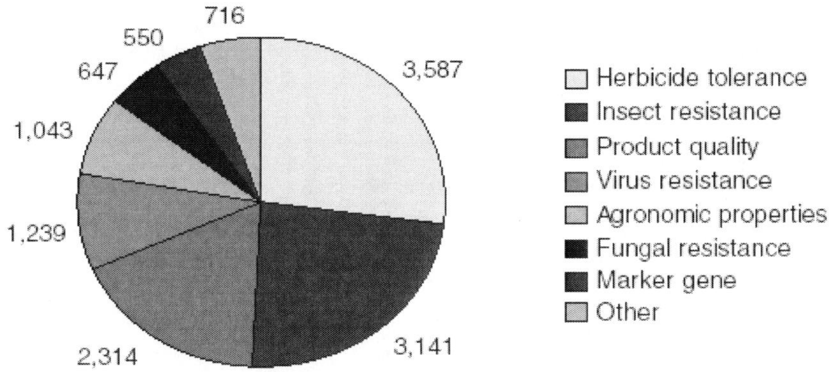

Source: Virginia Polytechnic Institute and State University, 2005.

Figure 3. Total number of permits approved by APHIS, by GE trait.

APHIS approvals for field testing also provide an indication of products that are in development and that may come "through the pipeline" in the future (table 2). In addition to crops with improved pest management traits, approvals include crops with traits that provide viral/fungal resistance, favorable agronomic properties (resistance to cold, drought, salinity, more efficient use of nitrogen), enhanced product quality (delayed ripening, increased protein and oil content, modified starch content, nutraceuticals (added vitamins, iron, antioxidants such as beta-carotene), and pharmaceuticals. Additional information may be found in Runge and Ryan and in Pew Initiative on Food and Biotechnology (2001).

After extensively field-testing a GE variety, an applicant may petition APHIS to deregulate (grant nonregulated status) the variety. If, after extensive review, APHIS determines that the new variety poses no significant risk to agriculture or the environment, permission is granted (see box, "Regulatory Oversight"). As of April 2005, APHIS had received 103 petitions for deregulation and had granted 63 (figure 4). Thirty-six percent of the released varieties have herbicide-tolerance traits, 27 percent have insect-resistance traits, and 17 percent have product-quality traits (figure 5).

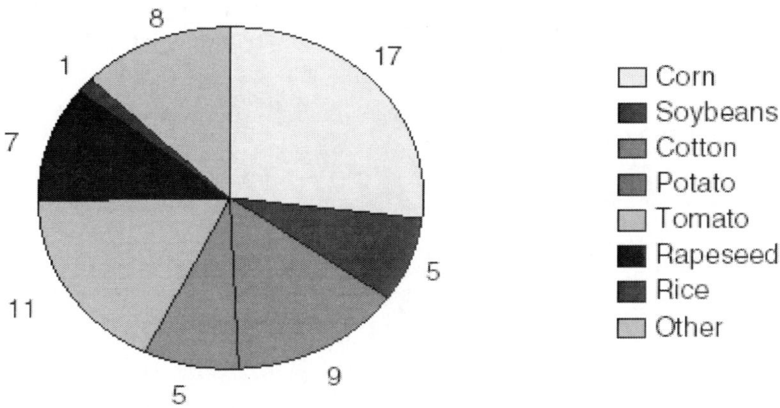

Source: Virginia Polytechnic Institute and State University, 2005.

Figure 4. Petitions for deregulation approved by APHIS, by crop.

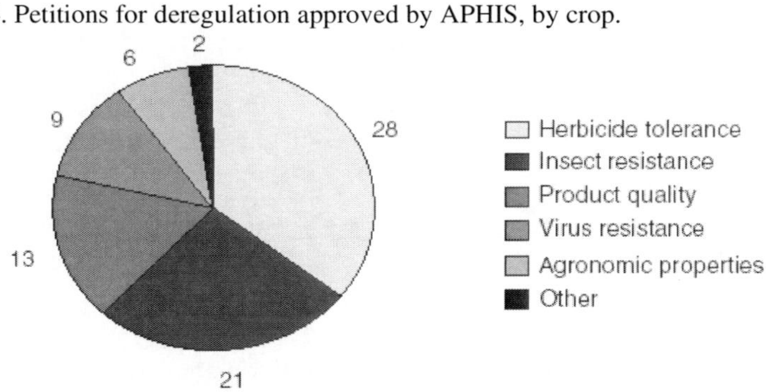

Source: Virginia Polytechnic Institute and State University, 2005.

Figure 5. Petitions for deregulation approved by APHIS, by GE trait.

Adoption of GE Crops by U.S. Farmers Increases Steadily

Farmers are more likely to adopt new practices and technologies if they expect to benefit from them. Benefits are usually thought of in monetary terms, but can also include ease of operation, time savings, lower exposure to chemicals, and other factors. Farmers choose technologies and practices they expect to yield the greatest benefit based on their own preferences, farm characteristics, demand for their product, and costs.

Farmers' expectations of higher yields, savings in management time, and lower pesticide costs have driven a rapid increase in the adoption of GE crop varieties in the United States and several other countries. An estimated 200 million acres of GE crops with herbicide tolerance and/or insect resistance traits were cultivated in 17 countries worldwide in 2004, a 20-percent increase over 2003. U.S. acreage accounts for 59 percent of this amount, followed by Argentina (20 percent), Canada and Brazil (6 percent each), and China (5 percent) (ISAAA, 2004).[3]

GE varieties of soybeans, corn, and cotton have been available commercially in the United States since 1996, and the rate of adoption by U.S. farmers has climbed in most years since then (figure 6). For the most part, farmers have adopted herbicide-tolerant (HT) varieties—which help control weeds by enabling crops to survive certain herbicides that previously would have destroyed them along with the targeted weeds—at a faster pace than insect-resistant (Bt) varieties.

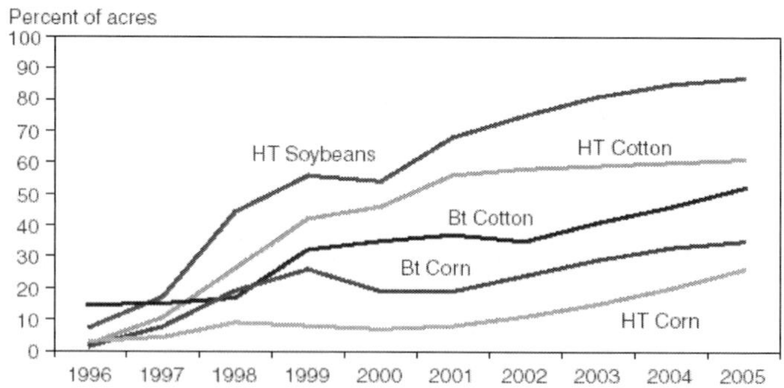

*Data for each crop category include varieties with both HT and Bt (stacked) traits.
Source: Fernandez-Cornejo (2005).

Figure 6. Adoption of genetically engineered crops grows steadily in the U.S.*

Weeds are such a pervasive pest for soybeans, corn, and cotton that over 90 percent of U.S. planted acreage for each crop has been treated with herbicides in recent years. The acreage share for HT soybeans has expanded more rapidly than that for HT varieties of cotton and corn, reaching 87 percent of U.S. soybean acreage in 2005.

Insect-resistant crops contain a gene from the soil bacterium *Bacillus thuringiensis* (Bt) that produces a protein toxic to specific insects. Acreage shares for Bt cotton and corn are lower than those for HT soybeans and cotton, and adoption is more concentrated in areas with a high level of infestation of targeted pests (insect infestation varies much more widely across locations than does weed infestation). Farmers planted Bt cotton to control tobacco budworm, bollworm, and pink bollworm on 52 percent of U.S. cotton acreage in 2005. Bt corn, originally developed to control the European corn borer, was planted on 35 percent of corn acreage in 2005, up from 24 percent in 2002. The recent increase in acreage share may be largely due to the commercial introduction in 2003/04 of a new Bt corn variety that is resistant to the corn rootworm, a pest that may be even more destructive to corn yield than the European corn borer (Comis).

Other GE crops planted by U.S. farmers over the past 10 years include HT canola, virus-resistant papaya, and virus-resistant squash (table 2). In addition, Bt potato varieties were introduced in 1996 but withdrawn from the market after the 2001 season, and a tomato variety genetically engineered to remain on the vine longer and ripen to full flavor after harvest was introduced in 1994 but was withdrawn from the market after being available sporadically for several years.

U.S. Farmers Expect to Profit from Adopting GE Crops

According to USDA's Agricultural and Resource Management Surveys (ARMS) conducted in 2001-03, most of the farmers adopting GE corn, cotton, and soybeans indicated that they did so mainly to increase yields through improved pest control (figure 7). Other popular reasons for adopting GE crops were to save management time and make other practices easier and to decrease pesticide costs. These results confirm other studies showing that expected profitability increases through higher yields and/or lower costs (operator labor, pesticides) positively influence the adoption of agricultural innovations.

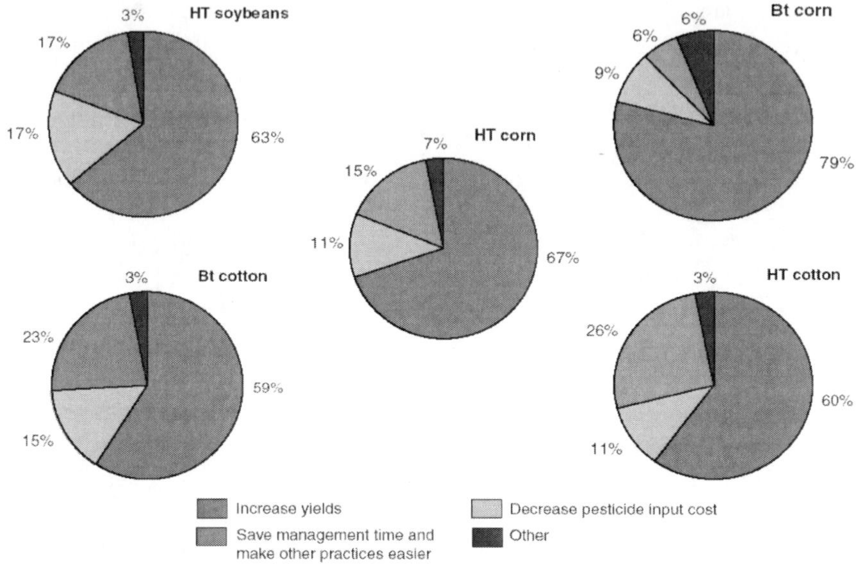

Source: Compiled by USDA's Economic Research Service using data from 2001, 2002, and 2003 Agricultural Resource Management Survey.

Figure 7. Farmers' reasons for adopting GE crops.

Adoption of GE Crops and Yields

Currently available GE crops do not increase the yield potential of a hybrid variety. In fact, yield may even decrease if the varieties used to carry the herbicide-tolerant or insect-resistant genes are not the highest yielding cultivars.[4] However, by protecting the plant from certain pests, GE crops can prevent yield losses compared with non-GE hybrids, particularly when pest infestation is high. This effect is particularly important for Bt crops. For example, before the commercial introduction of Bt corn in 1996, the European corn borer was only partially controlled using chemical insecticides. Chemical use was not always profitable, and timely application was difficult. Many farmers accepted yield losses rather than incur the expense and uncertainty of chemical control. For those farmers, the use of Bt corn resulted in yield gains rather than pesticide savings. On the other hand, a recently introduced Bt corn trait selected for resistance against the corn rootworm, previously controlled using chemical insecticides, may provide substantial insecticide savings.[5]

Many field tests and farm surveys have examined the yield and cost effects of using GE crops (table 3). The majority of the results show GE crops produce higher yields than conventional crops.

A 2002 ERS study found that increases in cotton yields in the Southeast were associated with the adoption of HT and Bt cotton in 1997—a 10- percent increase in HT cotton acreage led to a 1.7-percent increase in yield and a 10-percent increase of Bt cotton acreage led to a 2.1-percent increase in yield. Increases in soybean yields associated with the adoption of HT soybeans were statistically significant but small (Fernandez-Cornejo and McBride, 2002).[6]

A more recent ERS study using 2001 survey data found that, on average, actual corn yield was 12.5 bushels per acre higher for Bt corn than for conventional corn, an increase of 9 percent (Fernandez-Cornejo and Li, 2005).[7]

Adoption and Net Returns, Household Income, and Pesticide Use

The impacts of GE crop adoption on U.S. farmers vary by crop and technology. Many studies have assessed the effects of the adoption of GE crops on returns and pesticide use, and the results of these studies are summarized in table 3. ERS researchers found that:

Planting HT cotton and HT corn was associated with increased producer net returns, but HT corn acreage was limited.[8] The limited acreage on which HT corn has been adopted is likely to be acreage with the greatest comparative advantage for this technology. The positive financial association with adoption may also be due to low premiums for HT corn seed relative to conventional varieties in an attempt to expand market share (Fernandez-Cornejo and McBride, 2002).

Table 3. Summary of primary studies on the effects of genetically engineered crops on yields, pesticide use, and returns

Crop/researchers/ date of publication	Data source	Effects on		
		Yield	Pesticide use	Returns
Herbicide-tolerant soybeans				
Delannay et al., 1995	Experiments	Same	na	na
Roberts et al., 1998	Experiments	Increase	Decrease	Increase
Arnold et al., 1998	Experiments	Increase	na	Increase
Marra et al., 1998	Survey	Increase	Decrease	Increase

Table 3. (Continued)

Crop/researchers/ date of publication	Data source	Effects on		
		Yield	Pesticide use	Returns
Fernandez-Cornejo et al., 2002[1]	Survey	Small increase	Small increase	Same
McBride & El-Osta, 2002[2]	Survey	na	na	Same
Duffy, 2001	Survey	Small decrease	na	Same
Herbicide-tolerant cotton				
Vencill, 1996	Experiments	Same	na	na
Keeling et al., 1996	Experiments	Same	na	na
Goldman et al., 1998	Experiments	Same	na	na
Culpepper and York, 1998	Experiments	Same	Decrease	Same
Fernandez-Cornejo et al., 2000[1]	Survey	Increase	Same	Increase
Herbicide-tolerant corn				
Fernandez-Cornejo and Klotz-Ingram, 1998	Survey	Increase	Decrease	Same
McBride & El-Osta, 2002[2]	Survey	na	na	Increase
Bt cotton				
Stark, 1997	Survey	Increase	Decrease	Increase
Gibson et al., 1997	Survey	Increase	na	Increase
ReJesus et al., 1997	Experiments	Same	na	Increase
Bryant et al., 1999[3]	Experiments	Increase	na	Increase
Marra et al., 1998	Survey	Increase	Decrease	Increase
Fernandez-Cornejo et al., 2000[1]	Survey	Increase	Decrease	Increase
Bt corn				
Rice and Pilcher, 1998	Survey	Increase	Decrease	Depends on infestation
Marra et al., 1998	Survey	Increase	Decrease	Increase
Benbrook, 2001	Survey	Increase	na	Decrease
McBride & El-Osta, 2002[2]	Survey	na	na	Decrease
Duffy, 2001	Survey	Increase	na	Same
Pilcher et al., 2002	Survey	Increase	Decrease	na
Baute, Sears, and Schaafsma, 2002	Experiments	Increase	na	Depends on infestation
Dillehay et al., 2004[4]	Experiments	Increase	na	na
Fernandez-Cornejo & Li, 2005[5]	Survey	Increase	Decrease	na

na = not analyzed in the study.

[1] Results using 1997 data.

[2] Results using 1998 data.

[3] Results are for 1996 and 1998. Results were different in 1997 when pest pressure was low.

[4] Results using 2000-2002 data.

[5] Results using 2001 data.

[6] Net returns equal revenues minus variable costs.

Adoption of Bt cotton and corn was associated with increased returns when pest pressures were high. The adoption of Bt cotton had a positive association with producer net returns in 1997, but the association was negative for Bt corn in 1998. This suggests that Bt corn may have been used on some acreage where the (ex post) value of protection against the European corn borer was lower than the premium paid for the Bt seed. Because pest infestations vary from one region to another and from one year to another, the economic benefits of Bt corn are likely to be greatest where pest pressures are most severe. Farmers must decide to use Bt corn before they know what the European corn borer pest pressure will be that year, and damage caused by the European corn borer varies from year to year. Some farmers may have incorrectly forecast infestation levels, corn prices, and/or yield losses due to pest infestations, resulting in "overadoption." Also, producers may be willing to pay a premium for Bt corn because it reduces the risk of significant losses if higher-than-expected pest damage does occur (Fernandez-Cornejo and McBride, 2002).

Despite the rapid adoption of HT soybeans by U.S. farmers, no significant association with net farm returns was evident in 1997 or 1998. The lack of increased profitability for some farmers who adopted HT soybeans suggests that factors other than those included in traditional farm returns calculations may be driving adoption for these farmers. In particular, weed control may become simpler and require less management time, which allows growers of HT soybeans to control a wide range of weeds and makes harvest easier and faster. One important alternative use of management time is off-farm employment by farm operators and their spouses (Fernandez-Cornejo and McBride, 2002).

Adoption of HT soybeans is associated with increased household income. Recent ERS research showed that adoption of HT soybeans was associated with a significant increase in off-farm household income for U.S. soybean farmers. On-farm household income is not significantly associated with adoption but total farm household income is significantly higher for adopters, suggesting that most managerial time saved by adopters is used in off-farm work (Fernandez-Cornejo et al., 2005).

Adoption of GE crops is associated with reduced pesticide use. Pesticide use rates (in terms of active ingredient) on corn and soybeans have declined since the introduction of GE corn and soybeans in 1996 (figure 8). In addition, ERS research suggests that, controlling for other factors, pesticide use declined with adoption. There was an overall reduction in pesticide use associated with the increased adoption of GE crops (Bt and HT cotton, HT corn, and HT soybeans combined, using 1997/1998 data), resulting in a significant reduction in potential exposure to pesticides (Fernandez-Cornejo and McBride, 2002). Overall pesticide use on corn, soybeans, and cotton declined by about 2.5 million pounds, despite the slight increase in the amount of herbicides applied to soybeans. In addition, glyphosate used on HT crops is less than one-third as toxic to humans, and not as likely to persist in the environment as the herbicides it replaces (Fernandez-Cornejo and McBride, 2002).

More recently, using 2001 data, ERS found that insecticide use was 8 percent lower per planted acre for adopters of Bt corn than for nonadopters (Fernandez-Cornejo and Li, 2005).[9]

The ERS results generally agree with field-test and other farm surveys that have examined the effects of using GE crops (table 3). The majority of those results show that pesticide use for adopters of GE crops is lower than for users of conventional varieties.

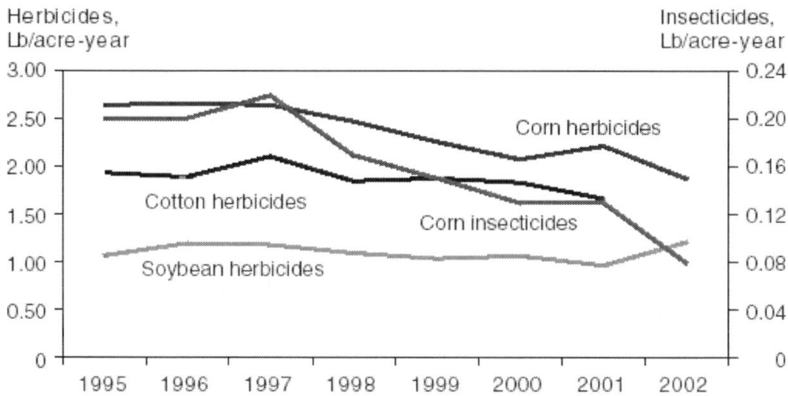

Source: NSS surveys.

Figure 8. Pesticide use in major field crops

Adoption of HT soybeans appears to be associated with conservation tillage. The environmental impact of conservation tillage is well documented.[10] The use of conservation tillage reduces soil erosion by wind and water, increases water retention, and reduces soil degradation and water and chemical runoff.

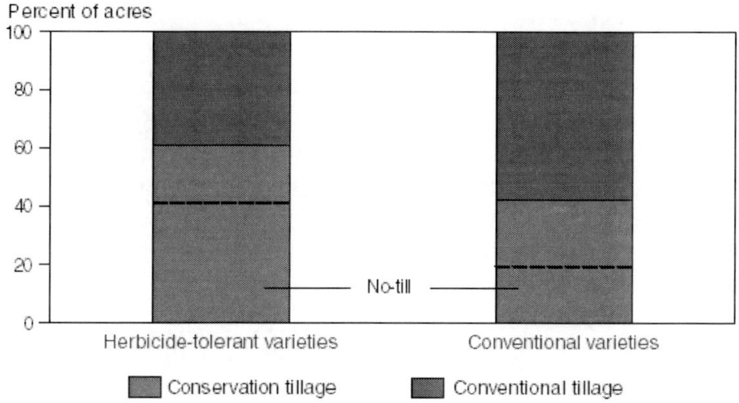

[*]Conservation tillage acres includes acres under no-till, ridge till, and mulch-till systems.

Source: Fernandez-Cornejo and McBride (2002).

Figure 9. Soybean area under conservation tillage[*] and no-till, 1997

According to USDA survey data, about 60 percent of the area planted with HT soybeans was under conservation tillage in 1997, compared with only about 40 percent of the acres planted with conventional soybeans (figure 9). Differences in the use of no-till between adopters and nonadopters of HT soybeans are even more pronounced: 40 percent of acres planted with HT soybeans were under no-till, twice the corresponding share of acreage planted with conventional soybeans. As a result, adoption of HT crops may indirectly benefit the environment by encouraging farmers to use soil conservation practices.

Consumer Demand Affects R&D, Adoption, and Marketing of GE-Derived Products

Investments in biotechnology-related research and development (R&D), the adoption of GE seeds, and the marketing of GE-derived products are all affected by consumer demand. While several surveys indicate that some U.S. consumers are concerned about GE food (table 4), these concerns have not had a large impact on the market for foods containing GE ingredients in the United States. In the European Union and a few other countries, consumer concern has resulted in substitution away from GE ingredients.

While opinion surveys give some indication of whether or not consumers are concerned about foods containing GE ingredients, they give little indication of the level of concern. Some researchers have attempted to quantify this concern through studies in which consumers are asked how much they would be willing to pay for foods made with GE ingredients, and for foods without GE ingredients. Researchers then use these data to measure whether or not there is a difference between these two hypothetical prices.

In most of these studies (table 5), consumers indicated that they were willing to pay more on average for GE-free foods or to avoid foods containing GE ingredients. However, in many of the studies, at least some consumers did not require a discount to buy foods containing GE ingredients, while some expressed that they would not be willing to buy foods containing GE ingredients at all.[11] Some respondents were willing to pay more for certain characteristics, such as improved nutrition and environmental benefits (Li et al., 2001; Lusk, 2003, Bocaletti and Moro, 2000).

Table 4. Surveys on Consumer Perceptions of Foods Containing GE Ingredients

Country/ Population	Surveyed by	Details
United States	Pew Initiative/ Mellman Group, 2003, 2004	27 percent favor introduction of GE foods;47 percent oppose. However, 64 percent disagree with the statement, "genetically modified foods should not be allowed to be sold even if the Food and Drug Administration believes they are safe," and 28 percent feel that those foods should not be allowed, even if the FDA feels they are safe.
United States	Gallup, 2001	52 percent support the application of biotechnology; 38 percent oppose the use of biotechnology in food production.
United States	Hallman, 2004	47 percent approved or leaned toward approval of the use of GE to make plant-based foods, 41 percent disapproved or leaned toward disapproval, and 12 percent were unsure.
United States	IFIC, 2005	50 percent said likely to buy and 45 percent said not likely to buy GE produce modified to taste better or fresher;64 percent said likely to buy and 32 percent said not likely to buy GE produce modified to require fewer pesticide applications.
Beijing, China	Hu and Chen, 2004	67 percent were concerned about biotechnology.
Nanjing, China	Zhong et al., 2002	40 percent would buy GE foods;17 percent would not; 34 percent don't know.

Country/ Population	Surveyed by	Details
Beijing, China, Shiajiazhuang, China	Ho and Vermeer, 2004	40 percent were willing or rather willing to consume foods containing GE-based ingredients, 51 percent were neutral, and 9 percent were rather unwilling or very unwilling to consume the foods.
Flemish speakers in Belgium	Verdurme and Viaene, 2003	15 percent opposed to GE foods;34 percent perceived small risks and small benefits;26 percent perceived moderate risks and moderate benefits; and 23 percent perceived large benefits.
United Kingdom	2003 GE Public Debate Steering board	86 percent preferred not to eat GE foods;8 percent happy to eat GE foods.

Source: Compiled by USDA's Economic Research Service.

While surveys and willingness-to-pay studies provide some insight into consumer opinion, they often do not reflect how consumers will behave in a real market situation when purchasing goods and services. Each food product has many characteristics, such as taste, color, and ripeness. The presence of a biotech-derived component is only one attribute. Empirically, it is difficult to determine what percentage of the price a consumer is paying for a specific characteristic. There are no published studies that indicate how many consumers have actually paid a premium to purchase non-GE goods, but there is some empirical evidence of the types of goods that are currently offered for sale to consumers. In the United States, many products contain GE ingredients, and the demands for these products apparently have been unaffected by negative opinions about biotechnology expressed in surveys. A few specialty brands are marketed as "GE free," but they represent a small percentage of supermarket sales.[12] In some other countries, however, strong consumer demand for non-GE products has limited the availability of GE items (see box, "Biotech Product Differentiation: A Tale of Two Markets").

Table 5. Willingness to Pay for Foods That do not Contain GE Ingredients[1]

Country	Food	Study	Willingness-to-pay premium
United States	Vegetable oil	Tegene et al., 2003	In experimental auctions, consumers willing to pay 14 percent more for non-GE food.
United States	Potatoes	Loureiro and Hine, ` 2002	Customers willing to pay 5 percent more for non-GE food.
United States	Golden rice	Lusk, 2003	Customers willing to pay 93 cents for GE "golden rice" with added vitamin C, 65-75 cents for regular rice.

Table 5. (Continued)

Country	Food	Study	Willingness-to-pay premium
United Kingdom	All foods	Burton et al., 2001	Customers indicated willingness to increase food budgets by 26-129 percent to avoid GE foods.
Italy	*	Bocaletti and Moro, 2000	Consumers willing to pay a positive amount for GE attributes;66 percent did not require a premium to consume GE foods.
United States, France, Germany, and United Kingdom	Beef fed with GE feed	Lusk et al., 2003	U.S.consumers willing to pay $2.83 and $3.31 per lb.to avoid GE; Euro-pean consumers $4.86 to $11.01.
United States, United Kingdom	Breakfast cereal	Moon and Balasubramanian, 2001	Survey found 56 percent of UK consumers willing to pay a premium to avoid GE compared with 37 percent of U.S. consumers.
Norway, United States, Japan, Taiwan	Vegetable oil	Chern et al., 2002	Norwegian students were willing to pay $1.51 (55-69 percent premium) per liter for non-GE vegetable oil, U.S. students were willing to pay $1.13 (50-62 percent premium), Japanese students were willing to pay $0.88 (33-40 percent premium), and Taiwanese students were willing to pay $0.45 cents (17-21 percent premium).
China	Rice	Li et al., 2002	80 percent of consumers did not require a premium to purchase GE rice and on average were willing to pay a 38-percent premium on GE rice and a 16-percent premium for GE soy oil.
Norway	Bread	Grimsrud, et al., 2004	Consumers required discounts of 37-63 percent to buy GE bread; One-fourth willing to buy with no discount.
Australia	Beer	Burton and Pearse, 2002	Younger consumers would pay $A 0.72 less and older consumers $A 0.40 less for beer made with GE barley.
Canada	*	West et al., 2002	83 percent of consumers ascribed a lower value to several GE foods.
France	*	Noussair et al., 2004	35 percent of consumers were unwilling to purchase GE foods, and 42 percent were willing to purchase them if they were less expensive.
United States	Oil, chips, and potatoes	Rousu et al., 2004	Consumers reduced their demand by an average of 7-13 percent for each food product having 1 percent and 5 percent tolerance levels for GE material relative to GE-free food.

[1] See also Lusk et al. (2005), who summarize a set of 25 studies including 57 GE valuation studies and report that, on average, consumers are willing to pay a positive premium for GE-free foods.

*This study did not focus on a specific food item.

Source: Compiled by USDA's Economic Research Service.

Adoption Offers Market Benefits to Many Stakeholders

In addition to farmers, seed suppliers, technology providers, and consumers also benefit from the adoption of GE crops in the United States. Biotechnology developers and seed firms benefit by charging technology fees and seed premiums to adopters of GE varieties. U.S. and foreign consumers may benefit indirectly from GE crops through lower commodity prices that result from increased supplies.[13] ERS estimated the total market benefit arising from the adoption of three GE crops in the United States—HT soybeans, Bt cotton, and HT cotton—in 1997 (Price et al., 2003).[14] Estimated benefits to farmers, seed producers, and consumers were around $210 million for Bt cotton, $230 million for HT cotton, and $310 million for HT soybeans. This estimate includes the change in total welfare in both the seed input and commodity output markets. The distribution of these benefits among consumers, farmers, technology providers (biotech firms), seed firms, and consumers and producers in the rest of the world (ROW) is shown in figures 10-12. The distribution of benefits varies by crop and technology because the economic incentives to farmers (crop prices and production costs), the payments to technology providers (biotech firms) and seed firms, and the effect of the technology on world crop prices are different for each crop and technology. For example, adoption of HT cotton benefits mainly consumers while Bt cotton benefits farmers and technology providers. Seed firms are by far the largest beneficiaries in the case of soybeans. These results should be interpreted carefully, since the estimates are based on only a few years of data. Moreover, estimated benefits and their distribution depend particularly on the analytical framework, supply and demand elasticity assumptions,[15] crops considered, and year-specific factors (such as weather). In particular, the benefits attributable to HT soybeans and their distribution are very dependent on the soybean supply elasticity. Table 6 shows estimates of the benefits of Bt cotton and HT soybeans and their distribution obtained by other researchers.

Biotech Product Differentiation: A Tale of Two Markets

The introduction of genetically engineered (GE) crops has led food manufacturers to make a choice for each of their products: either pursue a non-GE strategy and market and produce a non-GE product or source inputs based on cost and quality, or market and produce an undifferentiated product.

If all manufacturers were to pursue a non-GE strategy, farmers would eventually abandon GE technologies and consumer choice would be restricted to potentially higher cost non-GE products. If manufacturers were to pursue an undifferentiated strategy, then farmers' use of the technology would be determined by production costs and consumers would be faced with markets in which they could not differentiate between GE and non-GE foods. If manufacturers pursue both strategies, some farmers would continue to use the technology while others would grow conventional crops to supply non-GE markets. In this scenario, consumers would have a choice between GE and non-GE food, at least for some products.

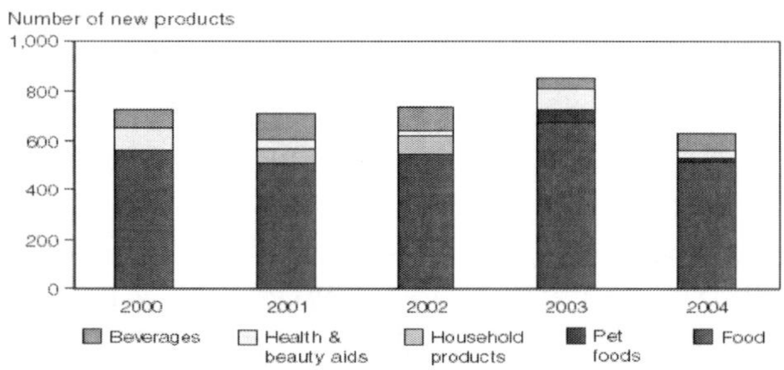

Source: Productscan Online.

Annual non-GE new product introductions in the United States.

In the United States, where unlabeled foods may contain GE ingredients, the data show that manufacturers have been active in creating a market for GE- free foods. From 2000 to 2004, manufacturers introduced over 3,500 products that had explicit non-GE labeling, mostly food products, with annual totals ranging from 854 in 2003 to 631 in 2004. This is in addition to organic foods (organic crops may not be grown using GE techniques) (Dimitri and Greene, 2002). In the European Union and Japan, where unlabeled foods cannot contain GE ingredients, manufacturers have chosen a non-GE marketing strategy. Very few products labeled as containing GE ingredients are found on European or Japanese grocery store shelves. The data also show that there have been limited attempts to market GE products in the United States. There were far fewer new GE products introduced than new non-GE products, and most of the GE products were introduced in the 1990s. GE products included tomatoes (advertised as better tasting with a longer shelf life), canola oil (advertised as heat stable), shrimp (advertised as gourmet-quality), beef (low-fat), dietary supplements, cigarettes (low-nicotine), and a drain cleaner.

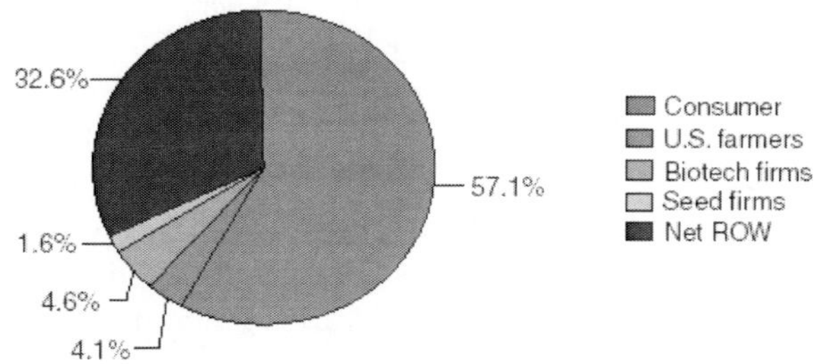

Source: Price et al., 2003.

Figure 10. Stakeholders' shares of the estimated total world benefit from adopting herbicide-tolerant cotton, 1997.

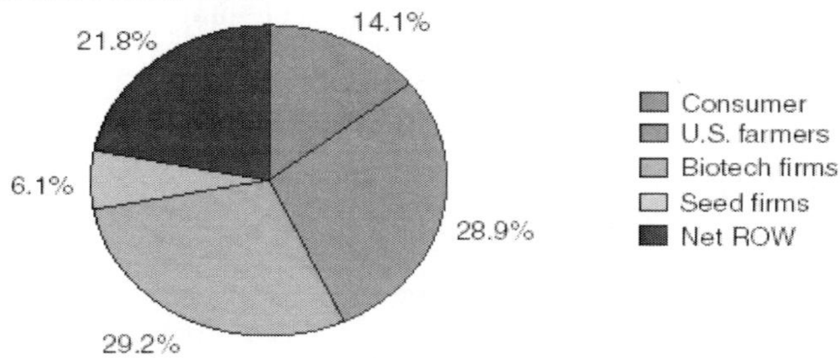

Source: Price et al., 2003.

Figure 11. Stakeholders' shares of the estimated total world benefit from adopting Bt-cotton, 1997.

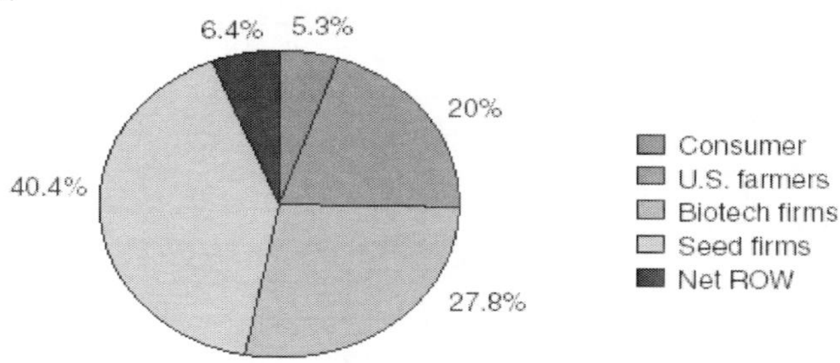

Source: Price et al., 2003.

Figure 12. Stakeholders' shares of the estimated total world benefit from adopting herbicide-tolerant soybeans, 1997.

CONCLUSION

The role that biotechnology plays in agriculture in the United States and globally depends on a number of factors and uncertainties. As the USDA Advisory Committee on Biotechnology and 21st Century Agriculture report indicates, "agricultural biotechnology sits at the crossroads of other debates on the future of American and world agriculture, on international trade relations, on biological diversity and the development of international instruments related to its preservation and exploitation, on the role of the multinational corporations, and on how best to build public confidence in rapidly evolving emerging technologies in general" (p.2.). One thing seems certain, however: the ultimate contribution of agricultural biotechnology will depend on our ability to identify and measure its potential benefits and its risks as well as their distribution.

Table 6. Benefits of GE Techniques and Their Distribution (from Estimates in Related Studies)

Study	Year	Total benefits	Share of the total benefits			
			U.S. farmers	Innovators	U.S. consumers	Net ROW
		$ million	*Percent*			
Bt cotton						
Falck-Zepeda et al. (1999)	1996	134	43	47	6	
Falck-Zepeda et al. (2000b)	1996	240	59	26	9	6
Falck-Zepeda et al. (2000a)	1997	190	43	44	7	6
Falck-Zepeda et al. (1999)	1998	213	46	43	7	4
Frisvold et al. (2000)	1996-98	131-164	5-6	46	33	18
EPA (2000)[1]	1996-99	16.2-45.9	n.a.	n.a.	n.a.	n.a.
Price et al. (2003)	1997	210	29	35	14	22
Herbicide-tolerant soybeans						
Falck-Zepeda et al. (2000a)	1997-Low elasticity[2]	1,100	77	10	4	9
	1997-High elasticity[3]	437	29	18	17	28
Moschini et al. (2000)	1999	804	20	45	10	26
Price et al. (2003)	1997	310	20	68	5	6

n.a. = Not applicable.

ROW = Rest of the world.

[1]Limited to U.S. farmers.

[2]Assumes a U.S. soybean supply elasticity of 0.22.
[3]Assumes a U.S. soybean supply elasticity of 0.92.
Source: Price et al., 2003.

REFERENCES

Arnold, J. C., Shaw, D. R. & Medlin, C.R. (1998). "Roundup Ready™ Programs Versus Conventional Programs: Efficacy, Varietal Performance, and Economics." *Proceedings of the Southern Weed Science Society, Southern Weed Science Society, 51*, 272-73.

Associated Press & Matt Moore. (2004). Covered by CBS news as "Europeans Balk At Biotech Beer" July 8, 2004. http://www.cbsnews.com /stories/2004/07/08/tech/main628305.shtml.

Baker, G. & Burnham, T. (2001). "Consumer Response to Genetically Modified Foods: Market Segment Analysis and Implications for Producers and Policy Makers," *Journal of Agricultural and Resource Economics, 26*(2), (December), 387-403.

Baute, T.S., Sears, M. K. & Schaafsma, A.W. (2002). "Use of Transgenic *Bacillus thuringiensis* Berliner Corn Hybrids to Determine the Direct Economic Impact of the European Corn Borer (*Lepidoptera: Crambidae*) on Field Corn in Eastern Canada. *Journal of Economic Entomology. 95*(1), 57-64.

Boccaletti, S. & Moro, D. (2000). "Consumer Willingness-To-Pay For GM Food Products In Italy," *AgBioForum, 3*(4), 259-267.

Bryant, K.J., W.C. Robertson, and G.M. Lorenz III. (1998). "Economic Evaluation of Bollgard Cotton in Arkansas." In *Proceedings of the Beltwide Cotton Conferences, National Cotton Council, Memphis, TN. 1*, 388-89.

Burton, M. & Pearse, D. (2002). "Consumer Attitudes Towards Genetic Modification, Functional Foods, and Microorganisms: A Choice Modeling Experiment for Beer." *AgBioForum, 5*(2), 51-58.

Burton, M., Rigby, D., Young, T. & James, S. (2001). "Consumer Attitudes to Genetically Modified Organisms in Food in the UK." *European Review of Agricultural Economics, 28* (4),(December), 479-98.

Carter, C.A., & Gruère, G. P. (2003(a)). "Mandatory Labeling of Genetically Modified Foods: Does it Really Provide Consumer Choice?" *AgBioForum, 6*(1 & 2), 68-70.

Carter, C. A., & Gruère, G. P. (2003(b)). "International Approaches to the Labeling of Genetically Modified Foods." *Choices*, June.

Chern, W. S., Rickertsen, K., Tsuboi, N. & Tsu-Tan Fu. (2002). "Consumer Acceptance and Willingness to Pay for Genetically Modified Vegetable Oil and Salmon: A Multiple-Country Assessment." *AgBioForum*, *5*(3), 105-112.

Colorado State University. (2004). Department of Soil and Crop Sciences. "Discontinued Transgenic Products," in *Transgenic Crops: An Introduction and Resource Guide*. http://www.colostate.edu/programs/life sciences/TransgenicCrops/ defunct.html. March.

Comis, D. (1997). "*Safe Corn Pest Bait Expected to Slash U.S. Insecticide Use*." U.S. Department of Agriculture, Agricultural Research Service. http://www.ars.usda.gov/is/pr/1997/970116.htm?pf=1.

Conservation Technology Information Center (CTIC). National Crop Residue Management Survey. "Conservation Tillage Data." http://www.ctic. purdue.edu/CTIC/CRM.html.

Culpepper, A.S. & York, A. C. (1998). "Weed Management in Glyphosate-Tolerant Cotton." *The Journal of Cotton Science, 4*,174-185.

Delannay, X., Bauman, T. T., Beighley, D. H., Buettner, M. J., Coble, H. D., DeFelice, M. S., Derting, C. W., Diedrick, T. J., Griffin, J. L., Hagood, E. S., Hancock, F. G., Hart, S. E., LaVallee, B. J., Loux, M. M., Lueschen, W. E., Matson, K. W., Moots, C. K., Murdock, E., Nickell, A. D., Owen, M. D. K., Paschall II, E. H., Prochaska, L. M., Raymond, P. J., Reynolds, D. B., Rhodes, W. K., Roeth, F. W., Sprankle, P. L., Tarochionc, L. J., Tinius, C. N., Walker, R.H., Wax, L. M., Weigelt, H. D. & Padgette. S. R. (1995). "Yield Evaluation of a GlyphosateTolerant Soybean Line After Treatment with Glyphosate." *Crop Science*, *35*, 1461-67.

Dillehay, B. L., Roth, G. W., Calvin, D. D., Karatochvil, R. J., Kuldau, G. A. & Hyde, J. A. "Performance of Bt Corn Hybrids, their Near Isolines, and Leading Corn Hybrids in Pennsylvania and Maryland." *Agronomy Journal*, *96*(204), 818-824.

Dimitri, C. & Greene, C. (2002). *Recent Growth Patterns in the U.S. Organic Foods Market.* U.S. Department of Agriculture, Economic Research Service. Agriculture Information Bulletin No. 777. Sept.

Duffy, M. (2001). "*Who Benefits from Biotehnology*?" Paper presented at the American Seed Trade Association meeting. Chicago, IL. December 5-7.

Falck-Zepeda, J. B., Traxler, G. & Nelson, R. G. (1999). "*Rent Creation and Distribution from the First Three Years of Planting Bt Cotton*." ISAAA

Briefs No. 14. International Service for the Acquisition of Agri-biotech Applications. Ithaca, NY.

Falck-Zepeda, J. B., Traxler, G. & Nelson, R. G. (2000a). "Rent Creation and Distribution from Biotechnology Innovations: The Case of Bt Cotton and Herbicide-Tolerant Soybeans in 1997," *Agribusiness, 16*, 21-32.

Falck-Zepeda, J. B., Traxler, G. & Nelson, R.G. (2000b). "Surplus Distribution from the Introduction of a Biotechnology Introduction." *American Journal of Agricultural Economics, 82,* 360-369.

Fernandez-Cornejo, J. (2004). *The Seed Industry in U.S. Agriculture: An Exploration of Data and Information on Crop Seed Markets, Regulation, Industry Structure, and Research and Development.* U.S. Department of Agriculture, Economic Research Service, Agriculture Information Bulletin No. 786. Feb.

Fernandez-Cornejo, J. (2005). "*Adoption of Genetically Engineered Crops in the U.S.*" Data product. U.S. Department of Agriculture, Economic Research Service. http://www.ers.usda.gov/Data/BiotechCrops/. July.

Fernandez-Cornejo, J. & Jiayi Li. (2005). "The Impacts of Adopting Genetically Engineered Crops in the USA: The Case of Bt Corn." Paper presented at the American Agricultural Economics Association meetings. Providence, RI..

Fernandez-Cornejo, J. & McBride, W. D. (2000). *Genetically Engineered Crops for Pest Management in U.S. Agriculture.* U.S. Department of Agriculture, Economic Research Service, Agricultural Economic Report No. 786. April.

Fernandez-Cornejo, J. & McBride, W. D. (2002). *The Adoption of Bioengineered Crops.* U.S. Department of Agriculture, Economic Research Service, Agricultural Economic Report No. 810. May.

Fernandez-Cornejo, J., Hendricks, C. & Mishra, A. (2005). "Technology Adoption and Off-Farm Household Income: The Case of Herbicide-Tolerant Soybeans." *Journal of Agricultural and Applied Economics, 37*(2).

Fernandez-Cornejo, J., Klotz-Ingram, C. & Jans, S. (2000). "Farm-Level Effects of Adopting Genetically Engineered Crops in the U.S.A.," In *Transitions in Agrobiotech: Economics of Strategy and Policy*, pp. 57-72. W.L. Lesser, editor. Food Marketing Research Center, University of Connecticut and Department of Research Economics, University of Massachusetts.

Fernandez-Cornejo, J., Klotz-Ingram, C. & Jans, S. (2002). "Farm-Level Effects of Adopting Herbicide-Tolerant Soybeans in the U.S.A." *Journal of Agricultural and Applied Economics*, *34*(1), 149-163.

Fernandez-Cornejo, J., Caswell, M. & Klotz-Ingram, C. (1999). "Seeds of Change: From Hybrids to Genetically Modified Foods." *Choices,* (Millennium issue, 4th quarter of): 18-22.

Finke, M. S. & Kim, H. (2003). "Attitudes About Genetically Modified Foods Among Korean and American College Students." *AgBioForum*, *6*(4), 19 1-197.

Frisvold, G., Tronstad, R. & Mortensen, J. (2000). "*Effects of Bt Cotton Adoption: Regional Differences and Commodity Program Effects*." Paper presented at the Western Agricultural Economics Association Meeting, Vancouver, British Columbia. June 30.

Gibson, J. W. IV, Laughlin, D., Lutrell, R. G., Parker, D., Reed, J. & Harris, A. (1997). "Comparison of Costs and Returns Associated with *Heliothis* Resistant Bt Cotton to Non-Resistant Varieties," *Proceedings of the Beltwide Cotton Conferences. Cotton Weed Science Research Conference. v.1,* pp. 244-247.

Goldman, D.L., Guy, C., McCelland, M. & Kendig, A. (1996). "Weed Control Systems in Roundup Ready^{TM} Cotton." In *Proceedings of the Beltwide Cotton Conferences. Cotton Weed Science Research Conference. v.2*. p. 1532.

Gracia, A. & Albisu, L. M. (2001). "Food Consumption in the European Union: Main Determinants and Country Differences." *Agribusiness, 17*(4),(Autumn), 469-488.

Graeber, J. V., Nafziger, E. D. & Mies, D.W. (1999). "Evaluation of Transgenic, Bt- Containing Corn Hybrids." *Journal of Production Agriculture, (12)(4)*, 659-663.

Greenpeace. EU Markets. (2005). http://www.greenpeace.org/raw/content /international/press/reports/ european-markets. February 3.

Grimsrud, K. M., McCluskey, J., Loureiro, M. & Wahl, T. (2004). "Consumer Attitudes to Genetically Modified Foods in Norway." *Journal of Agricultural Economics*, *55*(1),(March), 75-90.

Hallman, W., Hebden, W., Cuite, C., Aquino, H. & Lang, J. (2004). *Americans and GM Food: Knowledge, Opinion & Interest in 2004*. Food Policy Institute. http://www.foodpolicyinstitute.org/docs/reports/NationalStudy. pdf.

Harrison, R. W., Boccaletti, S. & House, L. (2004). "Risk Perceptions of Urban Italian and United States Consumers for Genetically Modified Foods." *AgBioForum*, *7*(4), 195-201.

Ho, P. & Vermeer, E. B. (2004). "Food Safety Concerns and Biotechnology: Consumers' Attitudes to Genetically Modified Products in Urban China." *AgBioFo rum*, *7*(4), 158-175.

House, L., Lusk, J., Jaeger, S., Traill, W. B., Moore, M., Valli, C., Morrow, B. & Yee, W. M. S. (2004). "Objective and Subjective Knowledge: Impacts on Consumer Demand for Genetically Modified Foods in the United States and the European Union." *AgBioForum*, *7*(3), 113-123.

Hu, W. Y. & Chen, K. (2004). "Can Chinese Consumers Be Persuaded? The Case of Genetically Modified Vegetable Oil." *AgBioForum*, 7(3), 24-132.

International Food Information Council (IFIC) Polls, 1999-2005. http://ific.org/publications/.

International Service for the Acquisition of Agri-biotech Applications (ISAAA). Preview: Global Status of Commercialized Biotech/GM Crops: ISAAA Briefs 32-2004. http://www.isaaa.org/kc/CBTNews/press_release/briefs32/ESummary/Executive%20Summary%20(English).pdf.

Just, R. E. & Hueth, D. L. (1993). "Multimarket Exploitation: The Case of Biotechnology and Chemicals." *American Journal of Agricultural Economics, 75*(4), 936-45.

Kaufman, P. (2000). "Consolidation in Food Retailing: Prospects for Consumers and Grocery Suppliers." *Agricultural Outlook*, August.

Keeling, W., Dotray, P., Jones, C. & Sunderland, S. (1996). "Roundup^{TM} Ready Cotton: A Potential New Weed Management Tool for the Texas High Plains." *Proceedings of the Beltwide Cotton Conferences. Cotton Weed Science Research Conference. v.2*. p. 1529.

King, J. & Heisey, P. (2003). "Ag Biotech Patents: Who is Doing What?" *Amber Waves*, November. http://www.ers.usda.gov/Amberwaves/November03/DataFeature/.

Li, Q., Curtis, K. R., McCluskey, J. J. & Wahl, T.I. (2002). "Consumer Attitudes Toward Genetically Modified Foods in Beijing, China." *AgBioForum*, *5*(4), 145-152.

List, J. A. & Gallet, C. (2001). "What Experimental Protocol Influence Disparities between Actual and Hypothetical Stated Values." *Environmental and Resource Economics, 20,* 241-254.

Loureiro, M. L. & Hine, S. (2002). "Discovering Niche Markets: A Comparison of Consumer Willingness to Pay for Local (Colorado

Grown), Organic, and GMO-Free Products." *Journal of Agricultural and Applied Economics, 34*(3), (December), 477-487.

Lusk, J. L. (2003). "Effects of Cheap Talk on Willingness to Pay for Golden Rice." *American Journal of Agricultural Economics*, *85*(4), (November), 840-856.

Lusk, J. L., Roosen, J. & Fox, J. (2003). "Demand for Beef with Growth Hormones and Fed Genetic Corn." *American Journal of Agricultural Economics*, *85* (1), (February), 16-29.

Lusk, J. L., Jamal, M., Kurlander, L., Roucan, M. & Taulman, L. (2005). "A Meta Analysis of Genetically Modified Food Valuation Studies." *Journal of Agricultural and Resource Economics*, *30*(April), 28-44.

Marks, L. A., Kalaitzandonakes, N. & Vickner, S. S. (2004). "Consumer purchasing behavior toward GM foods in the Netherlands," in *Consumer Acceptance of Biotech Foods*. R. Evenson and V. Santaniello (editors). Wallingford, UK: CABI Publishers, p. 23-29.

Marra, M., Carlson, G. & Hubbell, B. (1998). "Economic Impacts of the First Crop Biotechnologies." Available at http://www.ag.econ.ncsu.edu /faculty/marra/firstcrop/imp001.gif.

McBride, W. D. & El-Osta, H. (2002). "Impacts of the Adoption of Genetically Engineered Crops on Farm Financial Performance." *Journal of Agricultural and Applied Economics*, *Vol. 34*, No. 1(April), 175-19 1.

McBride, W., Short, S. & El-Osta, H. (2002). "*Production and Financial Impacts of the Adoption of Bovine Somatropin on U.S. Dairy Farms*." Paper presented at the 2002 American Agricultural Economics Association meetings, Long Beach, CA, July 28-3 1.

Moon, W. K. & Balasubramanian, S. K. (2001). "Public Perceptions And Willingness-To-Pay A Premium For Non-GM Foods in the US and UK." *AgBioForum*, *4*(3&4), 221-231.

Moon, W. K. & Balasubramanian, S. K. (2003). "Is There a Market for Genetically Modified Foods in Europe? Contingent Valuation of GM and Non-GM Breakfast Cereals in the United Kingdom." *AgBioForum*, *6*(3), 128-133.

Moschini, G., Lapan, H. & Sobolevsky, A. (2000). "Roundup Ready^{TM} Soybeans and Welfare Effects in the Soybean Complex." *Agribusiness, 16,* 3 3-55.

National Research Council. (2002). *Animal Biotechnology: Science Based Concerns*. Board on Agriculture and Natural Resources, Board on Life Sciences, Washington, DC: The National Academies Press. http://www.nap.edu/openbook/0309084393/html/R1.html#pagetop. (a).

National Research Council. (2002(b)). *Environmental Effects of Transgenic Plants: The Scope and Adequacy of Regulation*. Board on Agriculture and Natural Resources, Committee on Environmental Impacts Associated with Commercialization of Transgenic Plants. Washington, DC: The National Academies Press. http://www.nap.edu/books/0309082633/html/.

Noussair, C., Robin, S. & Ruffieux, B. (2004). "Do Consumers Really Refuse To Buy Genetically Modified Food?" *The Economic Journal*, *114*(492), 102-120.

North Carolina Farm Bureau Federation. (2001). "Gallup Conducts Poll on Biotech Food." Aug. 6-12.

Office of Science & Technology Policy. (1986). Executive Office of the President. *Coordinated Framework for Regulation of Biotechnology*. Federal Register, 51 FR 23302, http://usbiotechreg.nbii.gov/Coordinated_Framework_1986_Federal_ Register.html.

Onyango, B., Nayga, Jr. R. M. & Schilling, B. (2004). "Role of Product Benefits and Potential Risks in Consumer Acceptance of Genetically Modified Foods." *AgBioForum*, *7*(4), 202-211.

Panos. (1998). "Greed or Need? Genetically Modified Crops." Panos Media Briefing No. 30. October. http://www.oneworld.org/panos/.

Pew Initiative on Food & Biotechnology. (2001). *Harvest on the Horizon: Future Uses of Agricultural Biotechnology*. http://pewagbiotech.org /research/harvest/harvest.pdf. Sept..

Pew Initiative on Food & Biotechnology. (2003). "*Public Sentiment About Genetically Modified Food: September Update*." http://pewagbiotech.org /research/2003update/. 2003.

Pew Initiative on Food and Biotechnology. (2004). "Overview of Findings Focus Groups & Poll." http://pewagbiotech.org/research/2004update /overview.pdf. 2004.

Pilcher, C.D., Rice, M. E., Higgins, R. A., Steffey, K. L., Hellmich, R. L., Witkowski, J., Calvin, D., Ostlie, K. R. & Gray, M. (2002). "Biotechnology and the European Corn Borer: Measuring Historical Farmer Perceptions and Adoption of Transgenic Bt Corn as a Pest Management Strategy." *Journal of Economic Entomology, 95*(5), 878-892.

Price, G. K., Lin, W., Falck-Zepeda, J. B. & Fernandez-Cornejo, J. (2003). *The Size and Distribution of Market Benefits from Adopting Agricultural Biotechnology*. U.S. Department of Agriculture, Economic Research Service, Technical Bulletin No. 1906. Nov.

ReJesus, R. M., Greene, J. K. Hamming, M. D. & Curtis, C. E. (1997). "Economic Analysis of Insect Management Strategies for Transgenic Bt Cotton Production in South Carolina." *Proceedings of the Beltwide Cotton Conferences. Cotton Economics and Marketing Conference. v.1.* p. 247-251.

Rice, M. E. & Pilcher, C. D. (1998). "Potential Benefits and Limitations of Trans- genic Bt Corn for Management of the European Corn Borer (*Lepidoptera: Crambidae*)." *American Entomologist, 44,* 75-78.

Roberts, R. K., Pendergrass, R. & Hayes, R. M. (1998). "Farm-Level Economic Analysis of Roundup Ready^{TM} Soybeans." Paper presented at the Southern Agricultural Economics Association Meeting, Little Rock, AR, Feb. 1-4.

Rousu, M., Huffman, W. E., Shogren, J. F. & Tegene, A. (2004). "Are United States Consumers Tolerant of Genetically Modified Foods?" *Review of Agricultural Economics*, 26(February), 19-31.

Runge, C. Ford & Ryan, B. (2004). *The Global Diffusion of Plant Biotechnology: International Adoption and Research in 2004.* University of Minnesota. http://www.apec.umn.edu/faculty/frunge /globalbiotech04. pdf. Dec. 8.

Shoemaker, R., Harwood, J., Day-Rubenstein, K., Dunahay, T., Heisey, P., Hoffman, L., Klotz-Ingram, C., Lin, W., Mitchell, L., McBride, W. & Fernandez-Cornejo, J. (2001). *Economic Issues in Agricultural Biotechnology*. U.S. Department of Agriculture, Economic Research Service, Agriculture Information Bulletin No. 762. Feb.

Shoemaker, R., Johnson, D. & Golan, E. (2003). "Consumers and the Future of Biotech Foods in the United States." *Amber Waves*, November. http://www.ers.usda.gov/Amberwaves/November03/Features/futureof-biotech.htm.

Steering Board of the Public Debate on GM. *GM Nation?* (2003). *The Findings of the Public Debate.* http://www.gmnation.org.uk/docs /gmnation_finalreport.pdf.

Tegene, A., W. Huffman, M. Rousu, and J. Shogren. (2003). *The Effects of Information on Consumer Demand for Biotech Foods: Evidence from Experimental Auctions.* U.S. Department of Agriculture, Economic Research Service, Technical Bulletin No. 1903. April.

Traill, W. B., Jaeger, S. R., Yee, W. M. S., Valli, C., House, L. O., Lusk, J. L. , Moore, M. & Morrow, Jr. J. L. (2004). "Categories of GM Risk-Benefit Perceptions and Their Antecedents." *AgBioForum*, *7*(4), 176-186.

Tripp, R. (1998). "Regulatory Issues: Varietal Registration and Seed Quality Control." In Morris, M.L (editor), *Maize Seed Industries in Developing Countries*. Boulder, CO: CIMMYT (International Maize and Wheat Improvement Center). pp. 159-173.

United Nations Conference on Trade and Development (UNCTAD). (2005). *International Trade in GMOs and GM Products: A National and Multilateral Legal Framework*. Policy Issues in International Trade and Commodities Study Series No. 29. United Nations, New York, NY, and Geneva, Switzerland.

University of California. (1998). Davis Vegetable Research and Information Center, Pest-Resistance Management. "Insecticidal Bt-Potatoes." http://ucce.ucdavis.edu/files/filelibrary/5283/2196.pdf. Feb.

USDA Advisory Committee on Biotechnology and 21st Century Agriculture (AC21). (2005). *AC21 Consensus Report—Preparing for the Future*. Washington, DC. http://www.usda.gov/agencies/biotech/ac21/reports/scenarios-4-5- 05final.pdf.

U.S. Department of Agriculture. "Agricultural Biotechnology: Frequently Asked Questions. What Is Biotechnology?" http://www.usda.gov/agencies/biotech/faq/html.

U.S. Department of Agriculture. "Glossary of Agricultural Biotechnology Terms." http://www.usda.gov/wps/portal/!ut/p/_s.7_0_A/7_0_1OB?contentidonly=true&navid=AGRICULTURE&contentid=BiotechnologyGlosary.xml. October 2005.

U.S. Department of Agriculture, Animal and Plant Health Inspection Service. "*Changes in Regulation of the Products of Biotechnology by APHIS' Biotechnology Regulatory Services*." http://www.aphis.usda.gov/brs/eis/index.html. January 2004.

U.S. Department of Agriculture. (2005). Biotechnology, Animal and Plant Health Inspection Service, Regulatory Services. "United States Regulatory Oversight in Biotechnology Responsible Agencies – Overview." http://www.aphis.usda.gov/brs/usregs.html. April 5.

U.S. Environmental Protection Agency. (2000). "*Bacillus thuringiensis* Plant-Pesticides." Biopesticides Registration Action Document: Preliminary Risks and Assessments Sections.

U.S. Environmental Protection Agency. (2003). *A's Regulation of Biotechnology for Use in Pest Management*. http://www.epa.gov/pesticides/biopesticides/reg_of_biotech/eparegofbiotech.htm. June.

U.S. Food and Drug Administration. *Statement of Policy.* (1992). *Foods Derived from New Plant Varieties,* 57 FR 22984. http://vm.cfsan.fda.gov/~lrd/fr92529b.html. May.

U.S. Food and Drug Administration. (2005). Center for Food Safety and Applied Nutrition. *Biotechnology.* http://www.cfsan.fda.gov/~lrd /biotechm.html. May.

Vencill, W.K. (1996). "Weed Management Systems utilizing Herbicide-resistant Cotton." *Proceedings of the Beltwide Cotton Conferences. Cotton Weed Science Research Conference. v.2.* p. 1532-33.

Verdurme, A. & Viaene, J. (2003). "Consumer Beliefs and Attitude towards Genetically Modified Food: Basis for Segmentation and Implications for Communication." *Agribusiness, 19*(1),*(Winter)*, 91-113.

Virginia Polytechnic Institute & State University. (2005). "Information Systems for Biotechnology, Field Test Releases in the U.S." http://www.isb.vt.edu/cfdocs/fieldtests1.cfm. April 5.

West, G. E., Gendron, C., Larue, B. & Lambert, R. (2002). "Consumers' Valuation of Functional Properties of Foods: Results from a Canada-Wide Survey." *Canadian Journal of Agricultural Economics, 50*(4), *(December)*, 541-58.

Zhong, F., Marchant, M. A., Ding, Y. & Lu, K. (2003). "GM Foods: A Nanjing Case Study of Chinese Consumers' Awareness and Potential Attitudes." *AgBioForum, 5*(4), 136-144.

End Notes

[1] In the United States, under guidelines issued by USDA's Animal and Plant Health Inspection Service (as published in the *Federal Register*, 7CFR340: 340.1), genetic engineering is defined as "the genetic modification of organisms by recombinant DNA techniques" (Fernandez-Cornejo and McBride, 2000). A full biotechnology glossary is in USDA (2005).

[2] Some firms evolved in the 1990s toward developing "life sciences" complexes organized around the development of products such as agricultural chemicals, seeds, foods and food ingredients, and pharmaceuticals based on applications of related research in biotechnology and genetics. However, most of those life sciences companies have since divested their agricultural operations after "failing to realize adequate returns on their investments" (Shoemaker et al., 2003, p.32; Fernandez-Cornejo, 2004, p.42).

[3] Also, there has been an upward trend in the adoption of "stacked gene" varieties (with traits of herbicide tolerance and insect resistance) in the case of cotton and corn.

[4] This yield decrease occurred mostly in early years. HT or Bt genes were introduced into high-yielding cultivars in later years.

[5] Entomologists estimate that the corn rootworm causes up to $1 billion in corn yield losses and insecticide expenditures annually in the U.S. (Comis).

[6] The study used an econometric model that takes into consideration that farmers' adoption of GE crops and pesticide use decisions may be simultaneous and that farmers are not assigned randomly to the two groups (adopters and non- adopters) but that they make the adoption choices themselves. Therefore, adopters and nonadopters may be systematically different. Differences may manifest themselves in farm performance and could be confounded with differences due to adoption. This self- selectivity may bias the results, unless corrected. To account for simultaneity and self-selectivity, the model uses a two-stage econometric model.

[7] In addition, results using an econometric model with the 2001 data showed a small but statistically significant yield increase associated with farmers who adopted Bt corn relative to those using conventional corn varieties. (Fernandez-Cornejo and Li, 2005).

[8] Net returns equal revenues minus variable costs, which include pesticide and seed costs. Seed costs paid by adopters of GE varieties include a technology fee paid by farmers to biotechnology developers and premiums to seed firms.

[9] In addition, using an econometric model with the 2001 data, the ERS study showed a moderate but statistically significant insecticide reduction associated with farmers who adopted Bt corn relative to those using conventional corn varieties (a 4.11-percent decrease in insecticide use was associated with a 10-percent increase in Bt corn adoption).

[10] Conservation tillage includes any tillage and planting system that leaves at least 30 percent of the soil surface covered with crop residue. It includes no-till, ridge-till, and mulch-till (Conservation Technology Information Center, 2004).

[11] The amount that consumers indicate that they are willing to pay for a particular characteristic in a hypothetical situation is sometimes different from the amount that they actually pay when shopping (Lusk, 2003).

[12] In addition, organic foods are available. Use of any GE techniques bars a crop from being certified as organic. Although organic foods still have a small market share (1-2 percent) of total U.S. food sales, their sales have been rising at a rate of 20 percent annually (Dimitri and Greene, 2002).

[13] Consumers may also benefit directly when GE products of the second and third generation are commercialized.

[14] The study estimated the economic gains for various stakeholders associated with adoption by incorporating the potential yield enhancements and savings in pest control costs into models that derive each crop's supply shift resulting from biotechnology. Given domestic and export demands, counterfactual world prices and quantities demanded of the commodities—those that would have prevailed in the market if biotechnology had not been introduced—are determined from market equilibrium conditions. Producer and consumer surpluses in the U.S. and international markets and monopoly profits accruing to the biotech developers and seed firms are then calculated (Price et al., 2003).

[15] Elasticity measures the responsiveness of one economic variable to a change in another (e.g., price and quantity demanded). It is unit free and always expressed in percentage terms.

In: Genetically Engineered Crops in America ISBN: 978-1-63321-225-1
Editor: Fredrick G. Lawrence

Chapter 3

PRIVATE RESEARCH AND DEVELOPMENT FOR CROP GENETIC IMPROVEMENT*

Paul W. Heisey and Keith O. Fuglie

Seed[1] has been an essential input in crop production since the origins of agriculture, when farmers first began to save grains for replanting and to select seeds for desirable characteristics. The pace of crop genetic improvement accelerated early in the 20th century with the development of applied genetics and associated changes in plant breeding, seed production, and seed marketing. The development of the modern seed industry began about this time.

From the mid-1990s to the present, the private-sector seed industry has probably undergone more structural change than any of the other agricultural input industries covered in this study. Technological innovation in the form of modern, DNA-level biotechnology and changes in intellectual property rules have enabled private-sector companies to capture more value from the new seeds they develop. In the late 1990s, research intensity (R&D spending as a percentage of sales) in the seed industry accelerated past intensity in all other agricultural input sectors, and seed remains the most research-intensive sector to date.

* This is an edited, reformatted and augmented version of a study which originally appeared as chapter 2 in report number, ERR-130, *Research Investments and Market Structure in the Food Processing, Agricultural Input, and Biofuel Industries Worldwide*, issued by the U.S. Department of Agriculture, Economic Research Service, December 2011.

GLOBAL MARKET FOR CROP SEED

Seed by Sector of Origin

Seed used for crops has three main sources: farmer-saved or farmer-sourced seed, commercial seed from the public sector, and commercial seed from the private sector. Private-sector proprietary seed dominates markets globally today, particularly in high-income countries.

The dominance of private companies as the primary source of crop seed is a relatively recent phenomenon. Historically, farmers saved seed from their own crops or obtained seed from their neighbors for replanting in the next season. In some cases, poor crops or other negative factors may lead to consumption of saved seed, and after such periods, farmers may purchase seed from food markets to replenish their supplies. This is particularly common in some developing countries and for certain crops. Farmer-saved or "bin-run" seed continues to be a seed source, even for a major crop such as wheat, in high-income agricultural economies such as the United States. Plant variety protection laws, however, tend to restrict or forbid practices such as sales to other farmers.

As scientific plant breeding developed, public-sector breeders were often the major sources of new crop varieties. Public-sector varieties have sometimes been multiplied and sold to farmers by private seed distributors, but these distributors may not conduct seed-related research. In some developing countries, government-owned companies also distribute public seed varieties. As the seed industry develops further, however, private companies that perform their own plant breeding and seed research can become increasingly important (Morris et al., 1998). In high-income countries, where private-sector research on crop improvement is significant, public and private research efforts focus on complementary, but different, stages of the research process. For example, in plant breeding in the United States today, most public R&D is oriented toward basic breeding methods or basic germplasm enhancement, while most private R&D is devoted to commercial cultivar development (Frey, 1996). This balance between public and private-sector roles has continued to evolve over time, particularly with the more widespread commercial use of biotechnology techniques (Traxler, 1999). For some commodities (like wheat and potatoes in the United States), the public sector continues to provide most of the finished varieties because of a lack of private-sector interest (Heisey et al., 2001). Both public- and private-sector seed from scientific crop improvement programs are considered commercial seed, but in this study, we distinguish private-sector proprietary varieties from seed originating in the public sector.

Estimates of the size of the global commercial seed market in 2006 vary between $20 billion and $34 billion. Estimates of commercial seed value based on sales by companies that develop the seed may be somewhat lower than estimates based on farmer purchases. For purposes of historical comparison and disaggregation, we take an intermediate sales-based figure for commercial seed reported by the Context Network (2007), $22.9 billion. In addition, the value of farmer-saved seed in 2006 is estimated at $6.1 billion. Based on these two amounts, the total value of crop seed used in 2006 is estimated at $29 billion.[2] The $22.9 billion of commercial seed can be further subdivided into $19.6 billion of proprietary seed ($11.8 billion of conventional proprietary seed, $7.8 billion of GM proprietary seed, and $3.3 billion of public-sector seed) (table 2.1).

Table 2.1 shows estimates of the real value (in constant 2006 U.S. dollars) of the world seed market. Between 1995 and 2006, real market sales of commercial public sector seed declined, while sales of proprietary seed increased markedly. In all likelihood, this trend began well before 1995. In recent years, market sales of proprietary seed have been at least six times those of seed from the public sector.

Table 2.1. Size of the global seed market

Year	Proprietary conventional seed	Proprietary genetically modified seed	Total proprietary seed	Public commercial seed	Farmer-saved seed	Total value of all seed	ISF value of total seed
			Million constant 2006 U.S. dollars				
1995	13,447	95	13,542	5,550	6,333	25,425	
2001	11,847	3,645	15,492	3,539	5,923	24,954	34,173
2002	11,210	4,148	15,358	3,483	6,390	25,231	33,631
2003	11,084	4,938	16,022	3,409	6,694	26,125	32,922
2004	11,525	5,869	17,394	3,315	6,616	27,325	32,013
2005	12,082	6,815	18,897	3,408	6,402	28,707	30,979
2006	11,800	7,800	19,600	3,300	6,100	29,000	34,000

Sources: USDA, Economic Research Service using Context Network (2007) for all columns except ISF value of total seed, which is from International Seed Federation (ISF). Values adjusted for inflation by the U.S. Gross Domestic Product implicit price deflator *(Economic Report of the President,* 2009).

The first significant commercial sales of proprietary seed with GM traits occurred in 1995. Market sales of GM seed have increased rapidly; since 2006, they have exceeded 40 percent of the total sales value of proprietary seed. The real value of all seed has increased over the last 5 years, and this increase has been driven primarily by the expansion of the market for GM seed.

Seed Markets by Region and Commodity

Over the past decade, commercial seed markets in three broad world regions— Asia-Pacific, Europe/Middle East/Africa, and North American Free Trade Association (NAFTA)—have been roughly equal in market sales, although amounts vary from year to year. Seed markets in each region have generally fluctuated around 30 percent of the global total, as calculated from data reported by the International Seed Federation (ISF). The seed market in Latin America/Caribbean makes up the remaining 10 percent. In more recent years, the Middle East/Africa portion—under 5 percent of the global total—can be separated from that for Europe. Industry analysts tend to estimate a higher percentage for NAFTA and a lower percentage for Asia-Pacific, as indicated by recent (2010) data from the ISF.[3]

For 2006, the Context Network estimated that field crops accounted for 77 percent of the global market for proprietary seeds. This may be further broken down into grains (46 percent), oilseeds (20 percent), and other field crops (11 percent). Vegetable and flower seeds made up 14 percent of the proprietary seed market, and forage and turf grass 9 percent. We assume that these percentages apply to all commercial seed, including public-sector seed, in the absence of data disaggregating public-sector seed.

The Context Network also estimated the value of seed markets for major field crops. Including "technology values" for GM traits and royalties, corn constituted about 25 percent of the total global market for commercial seed between 2001 and 2005. The next largest commodity in terms of seed market sales was soybeans, with over 12 percent of the total. Both wheat—the largest crop worldwide in terms of acreage—and cotton constituted about 4 percent each of the global commercial seed market. The value of the seed market for these two crops is roughly equivalent, even though world wheat area is over six times larger than world cotton area (a much higher proportion of annual cotton seed requirements is sourced through commercial markets, and cotton seed unit prices are considerably higher in part because GM traits have been incorporated into cotton but not wheat). Rice, the second most widely grown crop worldwide, accounted for just over 1 percent of the commercial seed market (most annual rice seed requirements are sourced from farmer-saved seed). Apparent discrepancies between sales-based and farmer-purchase-based estimates may be larger for major crops such as wheat and rice, for which seed saving is widespread, than for such crops as corn, soybeans, or cotton, for which a larger percentage of seed used is purchased from seed companies. Furthermore, if the full market value of hybrid rice seed were accounted for, particularly for China where it is subsidized, the estimated rice share of global commercial seed might be higher.

The Market for Crop Genetic Traits

An alternative approach to partitioning the market, by GM traits, may be of economic interest but is more difficult to undertake with available data. Both large multinational companies that conduct research in agricultural biotechnology and small biotechnology firms develop products in two broad classes—GM traits that may be inserted into crop seed or research tools that facilitate biotechnology procedures, such as gene discovery or genetic modification. Determining licensing revenue that these firms obtain from traits or tools is generally not possible directly.[4] Large multinational companies include licensing revenue within their broadly defined "seed" category and will not separate it from actual sales of seed. Neither overall sales nor specific revenue from licensing are available for most small agricultural biotechnology firms, and in general, it is not possible to estimate licensing revenues for research tools.

The market value of technology fees and royalties, however, may provide a rough estimate of the value of GM traits, although it is not identical to this value. For some of the major crops, technology fees and royalties represent a significant proportion of the cost of seed. Globally, by 2005, these fees and royalties constituted about half the total market value of seed in the case of cotton and about a quarter in the case of soybeans. These are crops for which significant portions of world area (over 40 percent for cotton, and over 60 percent for soybeans) are planted to GM seeds (see James, 2007, for an estimate of the distribution of GM crops worldwide). By 2005, only about 10 percent of the value of corn seed came from technology fees and royalties. Historically, private-sector firms have been able to capture a greater return for breeding effort devoted to corn through the use of hybridization, which guarantees the purchase of seed every year (Morris et al., 1998). The use of GM corn, however, has increased rapidly in recent years (James, 2007), and so the percentage of corn seed value obtained from technology fees and royalties is still rising.

Public-sector wheat seed breeding has remained relatively more important worldwide than public-sector breeding in some other major crops because successful wheat hybrids have not been widely deployed and because other technical factors have not allowed private breeders to capture a greater share of the returns from wheat breeding. Nonetheless, royalties have made up about an eighth of total wheat seed values in recent years. Particularly in Europe, and more recently in other high-income countries, institutional arrangements have evolved in which seed distributors pay royalties from the sale of wheat seed to the breeders, private or public, of the varieties involved. Because GM wheat has not yet been planted commercially, royalties from the sale of wheat seed do not reflect the value of GM traits for this crop.

For the largest national seed market, the United States, it is possible to estimate the value of GM traits in corn, soybeans, and cotton using price data on "biotech" and "nonbiotech" seed from the USDA's National Agricultural Statistics Service (USDA/NASS),[5] USDA/NASS data on total crop area planted[6] and the share of total area planted to crops with GM traits,[7]and seeding rates from USDA's Agricultural Resource Management Survey (ARMS).[8] Using these data, we first calculated the value of germplasm for all seed used in a given crop, biotech and nonbiotech, by using the nonbiotech seed price. We then estimated the value of GM traits by applying the difference between the biotech and the nonbiotech price to the area planted to GM crops.[9] The estimates followed the same pattern as the global estimates, with trait value accounting for the greatest percentage of total seed value for cotton and the lowest percentage for corn. These differences among crops may reflect the degree of adoption of GM traits, the market value of GM traits in comparison to the market value of the underlying germplasm, or other market or technical factors that are less well understood. Biotech traits accounted for about half the total value of cotton seed in the United States in 2001. After continuing to rise for several more years, the share has remained at about 80 percent since 2007. Biotech traits represented only 7 percent of the U.S. corn seed market value in 2001 but rose to 28 percent in 2007 and to 37 percent in 2009. Biotech traits have fluctuated between 30 and 40 percent of the U.S. soybean seed market for much of the past decade, reaching a high of 42 percent in 2007 but falling to 30 percent in 2009. Based on these data, the value of biotech traits for corn, soybeans, and cotton taken together have represented 20 percent or more of the value of the entire U.S. seed market in recent years.

Several studies in recent years used national micro-level market data to analyze component pricing strategy for different biotechnology traits and market power for U.S. field crops for which GM varieties are important. For the case of corn (Shi et al., 2010a) and soybeans (Shi et al., 2009), these data can also be used as a check on the estimates based on the NASS data.[10] For cotton, Shi et al. (2010b) do not present data on national-level seed prices, but their data still allow for comparisons with the NASS-based estimates used in this study.[11] Nonetheless, calculations based on data from Shi et al. suggest the same relationships (e.g., the proportion of the value of the U.S. corn seed market attributable to GM traits is lowest among the three major crops with significant areas planted to GM varieties, and the proportion of the value of the U.S. cotton seed market made up of GM traits is the highest). Similar to our NASS-based calculations, calculations based on Shi et al.'s (2010a) data imply the trait percentage of value for corn has risen from 7 to 21 percent between 2001 and 2007, while the trait percentage of value for soybeans has been higher but more variable, estimated at around 35 percent in 2007.

STRUCTURE OF THE GLOBAL SEED INDUSTRY

Seed Company Market Structure in 2009

In this report, we divide the global crop seed and biotechnology industry into four tiers. The first tier is what we refer to as the "Big 6"—large multinational corporations with positions in both the markets for crop seed-biotechnology and agricultural chemicals. The second tier consists of other seed companies that do some research. This group includes companies that may have a significant global presence as well as smaller regional or local seed companies. A third tier consists of mostly small seed companies that do not conduct research themselves but only produce and sell seed under licensing or other commercial arrangements with the other companies or public-sector breeders who develop new varieties. Finally, a fourth tier comprises small and medium-size agricultural biotechnology companies. These companies do not generally sell seed but rather seek to commercialize a new genetic trait or biotechnology service or tool to other firms in the industry. Our survey identified more than 100 such "agricultural biotechnology startups" that have proliferated since the 1980s. Some of these companies have been bought out by larger companies and others have exited the market, but at least 30 startups were in operation as of 2008.

Table 2.2. Companies with over $100 million in crop seed and biotechnology sales in 2009, plus BASF

Company	Country of incorporation	Crop seed and biotech sales	Agricultural chemical sales	Nonagricultural chemical sales and R&D	Pharmaceutical sales and R&D	Agricultural biotechnology research
		Million U.S. dollars				
Monsanto	U.S.	7,297	3,527	Divested 1997	Divested 2000	>80% of crop R&D
DuPont/Pioneer	U.S.	4,806	2,320	Primary product	Divested 2001	>50% of crop R&D
Syngenta	Switzerland	2,564	8,491	Divested 1996	Divested 2000	>15% of crop R&D
Limagrain	France	1,370	0	--	--	>25% of crop R&D
KWS AG	Germany	996	0	--	--	Yes
Bayer	Germany	699	7,535	No	Human and animal health	>85% of crop R&D
Dow	U.S.	633	3,708	Primary product	Divested 1996	>85% of crop R&D
Sakata	Japan	485	0	--	--	Yes
Forage Genetics Int'l (Land O'Lakes)	U.S.	*412*	0	--	--	No
DLF-Trifolium	Denmark	*391*	0	--	--	Yes
Takii	Japan	*347*	0	--	--	Yes
Rijk Zwaan	Netherlands	265	0	--	--	Yes
In Vivo	France	217	0	--	--	Yes
BarenBrug Holland BV	Netherlands	208	0	--	--	Yes
Saaten-Union	Germany	*187*	0	--	--	Yes
RAGT Semences SA	France	*181*	0	--	--	Yes
Florimond Desprez	France	*162*	0	--	--	Yes
Euralis Group	France	154	0	--	--	Yes
Maisadour Semences	France	119	0	--	--	Yes
Stine Seeds	U.S.	*unknown*	0	--	--	Yes
BASF	Germany	*small*	5,065	Primary product	Divested 2000	100% of crop R&D

* Seed sales figures for Land O'Lakes refer to alfalfa/forage seed developed by Forage Genetics International. Land O'Lakes also distributes seed for other companies, such as Monsanto and Syngenta, but these sales are not included in the Land O'Lakes estimate.

Seed sales figures in italics are ERS estimates not derived directly from company data.

Sources: USDA, Economic Research Service using compiled company reports and press releases, Le Buanec (2007), Allison (2007), and PhillipsMcDougall.

Table 2.2 summarizes information on 21 companies that each had global seed sales of over $100 million in 2009, including BASF, which made significant investments in crop biotechnology but did not have substantial seed or trait sales. Six of these companies also had a significant market share for crop protection chemicals.

The Big 6 firms (Monsanto, Dupont, Syngenta, Bayer, Dow, and BASF) hold a unique position in integrating biological and chemical technologies in agricultural input markets. In 2009, three of these firms—Monsanto, DuPont/Pioneer, and Syngenta—were the top three global seed companies, and they also ranked fifth, sixth, and first, respectively, in global sales of crop protection chemicals.

Bayer, another Big 6 firm, only entered the seed market in 2002 with the acquisition of Aventis Crop Science; by 2009, it ranked sixth in global seed sales. Dow (fourth in crop protection sales in 2009) ranked seventh in global seed sales, and BASF (third in global crop protection sales), which began research in seed/biotechnology around 1998, reported no seed or trait sales as of 2009.[12] However, both BASF and Dow have made significant investments in agricultural biotechnology.

Two European-based companies, Limagrain and KWS, also had seed sales of over $600 million in 2009.

Limagrain, originally a producer-owned cooperative based in France, is active in both the markets for seeds of field crops and vegetables. KWS, based in Germany, concentrates on seeds of field crops. Limagrain and KWS have a joint venture, AgReliant, in the North American market.

In 2009, at least 13 companies had global seed sales between $100 million and $600 million. Three companies (Land O'Lakes/Forage Genetics International, DLF-Trifolium, and BarenBrug Holland BV) specialized in seed of forage and/or turf grass.

Another Dutch seed company, Rijk Zwaan, and two Japanese-based companies, Sakata and Takii, specialized in vegetable and/or flower seed. Several European-based companies, In Vivo, Saaten-Union, RAGT, Florimond Desprez, Euralis Group, and Maisador Semences, focused mainly on field crops but also had forage/turf grass products as well as sugar beets. Finally, the U.S. soybean breeding company, Stine Seeds, likely fits into this category as well, although Stine Seeds does not disclose sales information. In 2009, these companies (except BASF) invested in crop improvement and sold proprietary seed for a range of crops (fig. 2.1).

Company	Country	Maize	Wheat & small grains	Rice	Soybean	Canola/ rape/sun-flower	Cotton	Sugar-beet	Sugar-cane	Forage	Vegetables & flowers
Monsanto	U.S.	Y	Y		Y	Y	Y	*	Y	*	Y
DuPont/Pioneer	U.S.	Y	Y	Y	Y	Y	Y			Y	
Syngenta	Switzerland	Y	Y	*	Y	Y	*	Y			Y
Limagrain	France	Y	Y	Y	Y	Y				Y	Y
KWS AG	Germany	Y	Y					Y			
Bayer	Germany			Y		Y	Y				Y
Dow	U.S.	Y	*	*	Y	Y	Y			Y	
In Vivo	France		Y			Y				Y	
Saaten-Union	Germany	Y	Y			Y				Y	
Florimond Desprez	France		Y			Y		Y		Y	
Stine Seeds	U.S.	Y			Y						
RAGT Semences SA	France	Y	Y		Y	Y				Y	
Euralis Group	France	Y				Y					
Maisadour Semences	France	Y				Y					
Forage Genetics Int'l	U.S.									Y	
DLF-Trifolium	Denmark									Y	
BarenBrug Holland BV	Netherlands									Y	
Takii	Japan										Y
Sakata	Japan										Y
Rijk Zwaan	Netherlands										Y

Y = company has crop breeding and seed sales; * = company develops biotechnology traits and platforms only. Source: USDA, Economic Research Service using data compiled from company websites.

Figure 2.1. Crop R&D portfolios of leading seed companies.

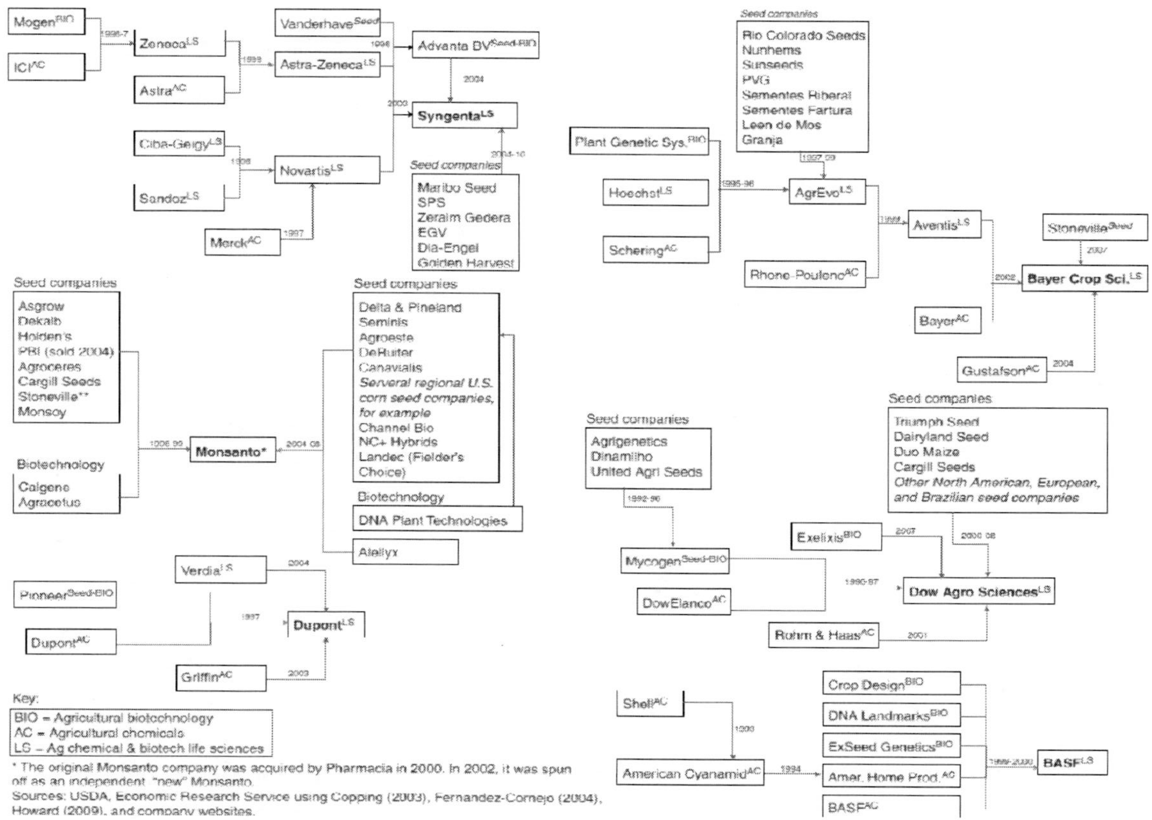

Figure 2.2. Formation of the "Big 6" seed-biotechnology-crop companies.

Changes in the Structure of the Global Seed Industry

Over the past 15 years, the seed industry has consolidated through mergers and acquisitions (see fig. 2.2 for global activity; Fernandez-Cornejo (2004) for U.S. activity; and Howard (2009) for graphics on current and historical seed industry ownership, including that of many small companies.[13]

Some of the features of changes to the seed industry can be summarized as follows:

1) Among the largest firms in terms of total product sales, the *close relationships between seed and agricultural chemicals industries* have continued. This applies to the Big 6 firms in particular (see fig. 2.2). These relationships may result partially from complementarity of product lines such as herbicide-tolerant seeds and chemical herbicides (Just and Hueth, 1993), or possibly from economies of scope in marketing as well.

Chemical companies also realized GM crops with pest resistance traits would compete with the crop protection chemicals, which helped drive these companies' interest first in biotechnology and eventually in seed, thus changing their business models to meet farmer demand for crop pest management as technological opportunities changed.

2) On the other hand, the *"life science industry" model* suggested a decade ago (Enriquez, 1998) *has not become the dominant paradigm.* This model stemmed from the likelihood that technologies underlying pharmaceutical discovery were the same as those underlying gene discovery for seeds.

Differences in business models and types of customer, however, prevented firms from combining both pharmaceuticals and agricultural biotechnology. Of the current Big 6 companies, only one— Bayer—has pharmaceuticals as its primary product line. Even when Bayer expanded into the seed/biotechnology industry in 2002 with its acquisition of Aventis Crop Science, Aventis pharmaceuticals eventually became a component of Sanofi-Aventis pharmaceuticals, not Bayer. Monsanto, which entered pharmaceuticals in the mid-1980s with its acquisition of Searle, was briefly held by Pharmacia before the agricultural enterprise was spun off as the "new Monsanto"; Pharmacia retained the pharmaceutical business segments. When Novartis's chemical and seed businesses were merged with Zeneca's agricultural chemical business in 2000 to form Syngenta, the pharmaceutical portion of Novartis remained intact as a separate large pharmaceutical company. BASF and DuPont ended their relatively smaller pharmaceutical investments after 2000 and 2001, respectively, and Dow had already sold its pharmaceutical subsidiary Marion Merrell Dow to Hoechst in 1996.

3) Agricultural chemicals have been an important part of product sales for all the Big 6 companies. However, positions in markets for nonagricultural chemicals have not remained constant, with some companies shedding these nonagricultural products. Monsanto divested this portion of its business to Solutia in 1997.

When Ciba-Geigy and Sandoz merged to form Novartis in 1996, nonagricultural chemicals were spun off to Ciba Specialty Chemicals, which eventually was acquired by BASF.

In response to antitrust considerations, Bayer sold selected insecticides and fungicides to BASF in 2003. DuPont sold its polymers business in the early 2000s.

Despite the common features in structural changes in the current Big 6, these and other former large multinational seed companies have followed somewhat individualized trajectories. Three of the current Big 6 were already identified as industry leaders 7 to 10 years ago: Monsanto, DuPont/Pioneer, and Syngenta (or Novartis) (Shoemaker et al., 2001; Fernandez-Cornejo, 2004). But several other potentially large players a decade ago no longer exist. Aventis Crop Science was acquired by Bayer in 2002, giving Bayer a position in the seed/biotechnology industry. Astra-Zeneca's seed business became part of the Advanta Seed Group, but Advanta's seed enterprise was broken up with parts acquired by Syngenta, Limagrain, and others in 2004 and 2005.

Monsanto has transformed itself most completely, from a chemical company to a seed/biotechnology company, and made by far the greatest number of large acquisitions of seed and related companies. Although Monsanto still has significant sales of the herbicide glyphosate, its research investments in chemicals are markedly reduced. Syngenta maintains a strong position in crop protection as well as in seed, but in contrast to its legacy companies, it is a wholly agricultural company. DuPont made by far the largest acquisition of all when it absorbed Pioneer, at the time the world's largest seed company, in 1999.[14] DuPont/Pioneer, however, has acquired only a few other seed or biotechnology firms since the merger. Dow was also recognized for its biotechnology investments a decade or more ago, particularly since it purchased Eli Lilly's share in Dow Elanco in 1997 and formed Dow Agrosciences. Though Dow Agrosciences has since acquired regional seed companies, a large portion of its investment in seed/biotechnology research has been in biotechnology. BASF has also focused almost completely on biotechnology, although for 10 years it had an alliance with the Swedish seed company Svalöf Weibull. Starting in 2007, BASF has concluded a number of research collaboration and licensing agreements with Monsanto and other seed companies for commercialization of future biotechnologies that BASF may develop. As noted, Bayer's entry into the seed business came with its acquisition of Aventis Crop Science. In general, as large chemical companies decided to commercialize their own biotechnology research or to buy seed company research, they needed to get access to seed companies either through direct acquisition, joint-venture, or licensing agreements. Through its purchase of Pioneer, DuPont gained immediate access to seed for multiple crops in multiple regions. Other companies had to acquire companies focusing on different crops in different regions; Monsanto has followed this strategy most comprehensively.

Many other large multinational seed companies have also followed a pattern of acquisitions or joint ventures. For example, Limagrain, which acquired other European seed companies starting with Vilmorin in 1975, acquired several more seed companies in the late 1990s and early 2000s, including Advanta Europe in 2005. Limagrain also formed a joint venture with KWS, AgReliant, in the North American corn and soybean markets, and also participated in projects with small biotechnology companies. DLF-Trifolium acquired the Dutch company Cebeco in 2002. Syngenta has held a significant minority interest in Maisadour for years. Recently, Svalöf Weibull (now Lantmännen SW Seed) and Florimond Desprez have initiated a research alliance, as have RAGT Semences and In Vivo.

Changes in Concentration in the Global Seed Market

In 2009, the top four companies accounted for 54 percent of the global commercial seed market (including public sector commercial seed), and the top eight companies accounted for 63 percent of total commercial seed sales. In 1994, these shares were 21 percent and 29 percent, respectively. Four-firm concentration ratios as measured by seed sales are now roughly equivalent across large agricultural input sectors. The eight-firm concentration ratio is still lower than eight-firm concentration in the agricultural chemicals industry (although four of the top six companies in each segment are the same), the animal health sector, and the animal genetics sector. Furthermore, the rate of growth of concentration in the seed market has been greater than the rate of growth for agricultural chemicals, animal health, or farm machinery.

The Herfindahl index (HI) is sometimes preferred to concentration ratios because it is sensitive to the distribution of market shares among firms, while the concentration ratios are not. In 1994, the HI for the global crop seed market was lower than similar indices for agricultural input sectors like animal health and agricultural chemicals. Between 1994 and 2009, HIs for seed, agricultural chemicals, animal health, and farm machinery all rose, but the rate of growth was most rapid for the seed sector. In 2009, the ranking for crop seed was higher than for any other sector, with the exception of animal genetics. Different sectors of the seed market can be more concentrated than the aggregate global seed market. For example, we estimate that the top four companies in the commercial market for vegetable seeds covered 70 percent of the global market in 2007, and the top eight companies 94 percent. Three of the Big 6 seed-chemical companies are major players in the global market for vegetable seeds—Monsanto since its acquisition of Seminis in 2005, Syngenta since its absorption of parts of the Advanta Seed Group in 2004, and Bayer since its acquisition of Aventis/Nunhems in 2002. Limagrain also sells a substantial amount of vegetable seeds and seeds for home gardens. Together with Sakata, Takii, and Rijk Zwaan, these companies represent the largest current participants in the vegetable seeds market.

Similarly, concentration in seed for particular crops in important individual markets can be higher than global concentration overall. For example, the four largest companies accounted for an estimated 72 percent of the U.S. market for corn (maize) seed in 2007 and 55 percent of the U.S. market for soybean seed (NRC, 2010). These shares have most likely continued to rise.[15] Data from USDA/Agricultural Marketing Service suggest that in 2009, the top two cotton seed companies, Bayer (Fibermax and Stoneville) and Monsanto (Deltapine), had 85 percent of the U.S. market for cotton seed, and the top four companies had 95 percent. These recent estimates compare with four-firm concentration ratios in 1998 of 67 percent for corn, 50 percent for soybeans, and 95 percent for cotton (Fernandez-Cornejo, 2004).

Agricultural Biotechnology and the Seed Industry

The structure of the seed industry has been transformed in large part because of the advent of modern agricultural biotechnology. Under the broadest definition, domestication of plant species and selection of desired characteristics within agricultural species would qualify as biotechnology. More narrowly, what *biotechnology* represents today is new knowledge about the natural processes of DNA replication, breakage, ligation, and repair that has made possible a deeper understanding of the mechanics of cell biology and the hereditary process itself (McCouch, 2001). Over the last 20-30 years, the term "biotechnology" in agriculture has been most closely associated with *genetic engineering*, but it may refer to a variety of techniques and products, including the use of molecular markers in genetic improvement or more general use of genomic information.

Although these modern techniques may appear to be simply additions to the genetic improvement toolkit, from an economic perspective, genetic engineering in particular has meant the development of three complementary markets: the traditional market for improved germplasm (or seed), the market for genes conferring traits that can be used to capture value, and the market for platform technologies or research tools. Intellectual property has become more prominent as a means of protecting traits, tools, and, to a certain extent, germplasm. Much of the merger and acquisition activity within the crop seed industry, as well as some research alliances and licensing agreements, is motivated by the desire to obtain access to products from all three of these markets, which are necessary for the final product of GM seed (Graff et al., 2003; King and Schimmelpfennig, 2005; and Marco and Rausser, 2008). Kalaitzandonakes and Bjornson (1997) argue that more well-defined and stronger intellectual property rights would have encouraged greater use of contracting and licensing arrangements and thus reduced the level of mergers and acquisitions.

In addition to intellectual property and R&D costs, costs of regulatory approval for agricultural biotechnology products constitute another type of fixed cost that can create barriers to market entry and thus influence industry concentration (Fulton and Giannakis, 2002). Kalaitzandonakes et al. (2006) estimate compliance costs for guiding a single genetic engineering event, for either insect-resistant corn or herbicide-tolerant corn, through the regulatory process at between $6 million and $15 million. Heisey and Schimmelpfennig (2006) argue that through the mid-2000s, the combination of desired characteristics and development of marketing and distribution networks were stronger determinants of industry concentration than regulation. However, it is likely that longer regulatory delays (Kalaitzandonakes et al., 2006) and regulatory costs for generic firms[16] as GM traits come off patent (Just, 2006) will have increasing effects on industry structure. Furthermore, the relative impact of regulatory costs might vary across crops as well as across countries

(Pray et al., 2006) and be significantly higher in many developing countries (Falck-Zepeda and Cohen, 2006).

Even given the increasing concentration of the seed industry with the advent of agricultural biotechnology, many different types of institutions may perform crop biotechnology research: public research institutions which may focus on more basic research (for example on research tools, in noncommercial model plants (e.g., *arabidopsis*) to understand gene function, or for crops which are not served by the private sector); small or medium-sized biotechnology companies that may or may not specialize in crop biotechnology; large seed companies; and integrated seed-chemical companies currently exemplified by the Big 6.

Small and Medium-Size Biotechnology Companies

Biotechnology companies specialize in research tools, identification and development of traits, or both. Many of these companies have been high-risk startups, depending on venture capital or "angel" investors, and turnover among these companies has been rapid. Significant entry into the marketplace began in the late 1970s and early 1980s, but in recent years, exits have outnumbered entrants. We identified 77 small and medium-sized companies[17] that have entered the agricultural biotechnology market since 1979 (fig. 2.3). All of these companies had agriculture as their primary business segment, and all but eight focused on crop biotechnology (these eight focused on animal biotechnology). Just over 30 companies are active as of 2008. Of the exits from the industry, about three-quarters were the result of acquisition and the remainder was due to bankruptcy or divestment.

In nearly all cases, we can only make estimates of the research investment by these firms, so the total estimated research expenditure by small biotechnology companies essentially provides the same information as a count of the number of active companies. In recent years, the total research investment of these companies is estimated to be about 5 percent or less of the total private-sector investment in seed/biotechnology research. Nonetheless, these companies developed some key agricultural biotechnology products, even though other companies eventually took the products to final market. For example, acquisitions of Agracetus and Calgene by Monsanto; Mogen by the Advanta group (now Syngenta); Mycogen by Dow; Plant Genetic Systems by AgrEvo (now Bayer, through Aventis); and DNA Plant Technologies by Seminis (now

Monsanto) have all contributed significantly to the eventual holders' portfolios of traits and research tools (Graff et al., 2003).

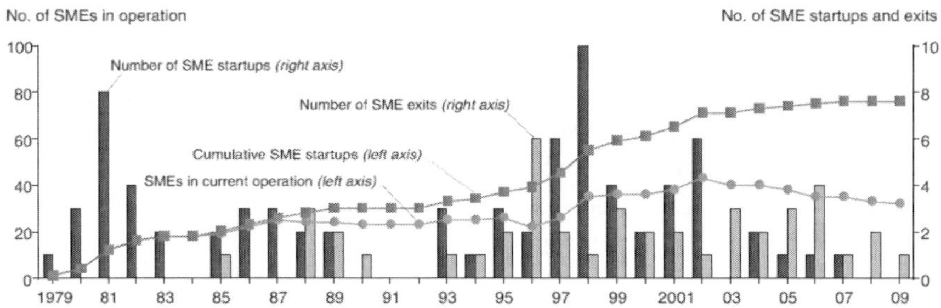

Source: USDA, Economic Research Service using data compiled from company websites and print media. Includes data on 77 small and medium enterprises (SMEs) for which agricultural biotechnology was or is the major focus.

Figure 2.3. Small and medium agricultural biotechnology startups and exits since 1979.

Large Seed Companies and the Big 6

All of the seed companies listed in table 2.2 have instituted biotechnology research, acquired interest in biotechnology companies, collaborated in biotechnology research, or signed licensing agreements for biotechnology products. In many cases, they have combined more than one of these activities. In several instances, seed companies that have not initiated GM research have still made use of marker-assisted breeding.[18] While the fixed cost of establishing in-house capacity in biotechnology may be prohibitive for small or midsized seed companies, strategic partnerships between firms can enable such firms to access biotechnology. For example, a number of midsized vegetable breeding companies established a joint venture, Keygene, to conduct biotechnology research on their commodities of interest. Significant cross-licensing agreements still exist between companies, including those comprising the Big 6 (Howard, 2009). For example, Monsanto has cross-licensing agreements with all the other Big 6 companies; Dow with four of the other five, and DuPont and Syngenta with three of the other companies. Despite the increase in strategic partnerships and research collaboration, in terms of both current biotechnology research expenditures and current control of GM traits, only a few large companies dominate the market.

TRENDS IN PRIVATE R&D INVESTMENT AND INNOVATION IN CROP GENETIC IMPROVEMENT

Research Spending by Private Seed and Crop Biotechnology Companies

Globally, real private sector research expenditures on crop seed development and crop biotechnology have risen substantially since the mid-1990s (table 2.3).

This increase can be decomposed in a number of ways. First, research intensity for seed increased over much the 1990s. Since 2000, research intensity has fallen in the seed industry, although it is still higher than that for animal health, animal genetics, or agricultural chemicals, the other agricultural input industries with high research intensities. But over most years since 2000, the real value of the global seed market has increased (see table 2.1), and real research expenditures have continued to grow. The slight decline in research intensity in recent years has been the result of sales increasing even more rapidly than research investments.

Structurally, these increases in research expenditures have been marked by the remarkable rise in the share accounted for by integrated seed-chemical companies, characterized by the Big 6.[19] The share of total private-sector crop seed and biotechnology R&D spending by seed-chemical companies began to grow with the first round of mergers and acquisitions in the mid-1990s, and it surpassed the share of other large seed companies in 1999, the year that DuPont acquired Pioneer.

In 2007, seed-chemical companies accounted for over 70 percent of total global R&D spending by the seed-biotechnology industry. Other seed companies made up an additional 24 percent, while small and midsized biotechnology firms accounted for 4 percent of R&D spending by this industry.

The average research intensity of the largest seed companies is also higher than that of midsized and small seed companies, but small biotechnology startups as a group account for the highest research intensity of all, at over 40 percent in 2006.

Table 2.3. Private-sector expenditures on crop seed and biotechnology research and development (R&D)

Year	Agricultural seed-chemical companies[1]	Other seed companies	Small and medium biotechnology firms	Total private seed and biotechnology R&D
		Million constant 2006 U.S. dollars		
1994	320	976	166	1,462
1995	355	1,013	168	1,536
1996	292	1,149	202	1,643
1997	576	1,139	144	1,859
1998	786	1,164	128	2,078
1999	1,091	908	128	2,127
2000	1,411	857	126	2,394
2001	1,312	842	141	2,295
2002	1,239	826	150	2,215
2003	1,266	857	142	2,265
2004	1,394	802	131	2,326
2005	1,453	746	129	2,328
2006	1,574	691	108	2,374
2007	1,764	676	100	2,540
2008	2,157	691	94	2,941
2009	2,353	702	93	3,149
2010	2,653	732	92	3,477

[1]These are companies with sales and R&D in both crop protection chemicals and seed/biotechnology. Since 2002 this group has been composed of the "Big 6" (BASF, Bayer, Dow, Dupont, Monsanto and Syngenta). Previously, AgrEvo, Astra-Zeneca, Aventis, Ciba-Geigy, Novartis and Sandoz were also part of this group.

Sources: USDA, Economic Research Service using data compiled from company reports, ASGROW (2007), Duncan (2007), author interviews with selected companies, and author extrapolations as described in text. R&D expenditures adjusted for inflation by the U.S. Gross Domestic Product implicit price deflator (*Economic Report of the President, 2009*).

Regionally, firms based in NAFTA countries, particularly the United States, and in Europe/Middle East/Africa (primarily Europe) dominate private research spending in crop seed and biotechnology. In 2006, over 53 percent of total investment was made by U.S.-based firms and another 42 percent was made by European-based firms. About 4 percent was attributed to Asia-Pacific firms and less than 1 percent to firms based in Latin America. With the globalization of the seed industry, however, many large firms are spending research dollars in other than their home regions.

It is not easy to estimate how much research is devoted to any particular crop. Companies breeding multiple crops may share resources, such as biotechnology-enabling technologies, across crops or may have other expenses, such as the costs of regulatory compliance, that are difficult to allocate. Nonetheless, it is clear that the largest proportion of seed/biotech research is directed toward corn (maize). Expert opinion also suggests that corn accounts for about 45 percent of all private-sector seed-related research (Cavalieri, 2009). This share is considerably higher than corn seed's 25-percent share of the overall seed market, but it is in accordance with industry estimates that corn seed is the most profitable seed for private companies to produce. Besides sales, profitability would be expected to contribute to relative seed research investment. We do not have enough information to estimate how much research is conducted on other individual crops, but we hypothesize that soybeans, cotton, and wheat research all command higher proportions of the private-sector total than their respective shares of the global seed market.

It is also difficult to estimate how much private R&D is devoted to conventional breeding research versus biotechnology. Le Buanec (2007) provides estimates of the relative proportions of seed-related research using biotechnology (including both genetic modification and marker-assisted selection) for several major companies in 2003. Using these estimates, and based on the assumptions that (1) 5 percent of other large seed companies' (see table 2.2) research is allocated to biotechnology, (2) 1 percent of other seed companies' research is allocated to biotechnology, and (3) all of the research for dedicated crop biotechnology companies is, indeed, allocated to biotechnology, we estimate that about half of the total research expenditures by the private seed industry in 2003 were allocated to biotechnology-related endeavors.[20]

Ownership Concentration in Crop Biotechnology Innovations

While data on research expenditures provide some indication of investments in innovation, they do not provide much information on the amount of new innovation actually taking place or its ownership. To gauge the sources of technological innovation emanating from private R&D, we assembled a number of indicators of intermediate research outputs and assigned them to the company or institution producing them. Our choice of indicators was necessarily limited by data availability.

For example, we did not obtain counts of new crop varieties released (a reliable indicator of crop research output) due to the difficulty in compiling a relatively complete dataset with wide global coverage from publicly available sources.[21]

The most complete data in this case might be that for European varietal registration, but linking named varieties with holders of plant variety protection rights would be extremely difficult. In the world's largest seed market, the United States, data are available for varieties with plant varietal protection certificates, varieties with utility patents, and varieties submitted for cultivar registration, but these data are both overlapping and nonuniversal in coverage.

The indicators reported here include the number of (1) agricultural biotechnology patents[22] issued in the United States, (2) petitions and notifications to USDA's Animal and Plant Health Inspection Service (USDA/APHIS) to import, transport across State lines, or release into the environment GM crops in the United States, (3) global approvals for planting or environmental release of GM crops, and (4) acres (multiplied by the number of traits per acre to get "trait-acres") planted to GM crops worldwide. These indicators are undoubtedly influenced by research investments, but the relationships are likely to be nonlinear and may change over time.

We summarize these data by showing the shares of select output held by a particular company or institution (see table 2.4). Those shares assigned to the eight listed seed companies include assignments to their historical counterparts, acquisitions (e.g., "legacy companies"), and subsidiaries. The counts for each indicator are cumulative (summed over several years) and therefore indicative of the overall dominance of the firm or institution in the market for seed and biotechnology innovations. But the indicators do not reflect changes in the relative market position of the firms or institutions over time. For example, patent counts are the total issued to a firm between 1982 and 2007 in the case of patents for crop cultivars, and between 1976 and 2000 for other classes of agricultural biotechnology patents (described below). The counts of USDA/APHIS petitions and notifications are the number issued since 1985, when USDA/APHIS first began issuing permits regulating the use and movement of GM seed, through mid-2008.[23]

Table 2.4. Shares of selected research outputs held by major seed companies and other institutions

Company (including subsidiaries and acquisitions	U.S. patents issued for crop cultivars	U.S. patents issued for agricultural biotechnology, narrowly defined[1]	U.S. patents issued for agricultural biotechnology, broadly defined[1]	Petitions and notifications to APHIS for field trials with GM plants in the U.S.	Approvals for planting or environmental release of GM crops	GM trait-acres of soybeans, corn, cotton, canola[2]
	% of total issued 1982 to 2007	% of total issued 1976 to 2000	% of total issued 1976 to 2001	% of total field trials 1985 to mid-2008	% of global approvals 1985 to 2007	% of global trait-acres in 2007
Monsanto	28.7	16.8	6.1	39.7	49.2	about 85
Dupont	36.5	20.7	5.1	7.3	5.0	3-5
Syngenta	8.4	9.8	3.8	5.2	8.0	3-5
Bayer	0.2	5.8	4.7	4.9	20.9	3-5
Dow	2.3	9.9	2.1	3.5	4.0	3-5
BASF	0.0	0.9	2.1	1.0	0.0	0
Limagrain	1.2	0.5	n.a.	1.2	0	0
KWS	0.4	0.1	n.a.	1	<0.1	0
Other private firms or	21.7	15.1	43.8	9.8		
individuals					12.9	2
Public and nonprofit institutions	0.6	20.4	32.3	26.4		
Total %	100.0	100.0	100.0	100.0	100.0	100.0
Total number				38,978	306	352 (282)[3]
						(mil. acres)

n.a. = not available. GM = genetically modified crops.

[1]The ERS Agricultural Biotechnology Intellectual Property (ABIP) database defines agricultural biotechnology two ways. Under a narrow definition, it includes patents pertaining specifically to crops and to the suite of modern biotechnology techniques, such as genetic engineering, selection with the aid of molecular markers, or genomics. Under the broader definition, biotechnology refers to any "use of organisms or parts of an organism to make or improve products or processes" in food or agriculture. See King and Schimmelpfennig (2005) for further information regarding ABIP.

[2] A "trait-acre" is the area sown to GM crops, where stacked GM traits are counted as multiple acres, depending on the number of traits stacked in a single seed. Total trait-acres include data from 13 countries with at least 100,000 ha in total GM crop area in 2007.

[3]The first figure is for total GM trait-acres. The second figure is the total area planted to GM crops in 2007 (James, 2007).

Sources: USDA, Economic Research Service using the following: U.S. patents issued from crop cultivars from the U.S. Patent and Trademarks Office; U.S. patents issued for agricultural biotechnology (narrow and broad definitions) from the ERS Agricultural Intellectual Property database; petitions and notifications to APHIS are from Virginia Polytechnic Institute and State University; approvals for plantings or environmental release of GM crops and GM trait-acres based on data provided in James (2007).

Agricultural patents are counted three ways. The first column of table 2.4 is based on U.S. patents granted for crop cultivars through 2007. Three firms accounted for nearly three-fourths of these patents; these operations are the three largest seed companies in both the United States and the world. Ninety percent of these patents were issued for two crops: corn (either inbred parent lines or hybrids) and soybeans, with cotton a distant third with just over 2 percent of total patents. One midsized U.S. seed firm, Stine Seeds (not listed separately in the table), held nearly 13 percent of the total soybean cultivar patents. As might be expected, public or nonprofit institutions held few cultivar patents.

The second and third columns in the table summarize data from ERS's Agricultural Biotechnology Intellectual Property (ABIP) database. The ABIP database includes all agriculturally related biotechnology patents granted between 1976 and 2000. The data in the second column use a narrow definition for agricultural biotechnology, namely, patents that pertained specifically to crops and to the use of "modern" biotechnological techniques such as genetic engineering, selection with the aid of molecular markers, or genomics. The data in the third column are based on a broader definition of agricultural biotechnology, where biotechnology refers to any "use of organisms or parts of an organism to make or improve products or processes" in agriculture or food production. Both the narrow and broad patent definitions include patents issued for crop cultivars (the patents counted in the first column) through 2000. The Big 6 companies account for nearly two-thirds of the modern crop biotechnology patents (narrow definition) and one-quarter of the more broadly defined agricultural biotechnology patents. As with cultivar patents, the majority of these patents were issued in the names of legacy companies or subsidiaries. Unlike with cultivar patents, nonprofit institutions hold a notable minority of these biotechnology patents. Unfortunately, categorizing biotechnology patents is difficult and subject to error, and data coverage extends only through the end of the year 2000. Thus, patenting by more recent entrants may be underreported.

The fourth column presents data on petitions and notifications to USDA/ APHIS on the importation, interstate transport, and environmental releases of GM seed. The Big 6 firms accounted for 62 percent of these petitions and notifications, with Monsanto claiming nearly 40 percent of the total.

In this instance, more petitions were recorded in the name of the parent firm than in the names of legacy companies. Public and nonprofit institutions accounted for about one-quarter of these petitions.

The final two columns are based on data on global use of GM seed reported by James (2007). The second-to-last column summarizes approvals for planting and/or environmental release of a specific GM "event" through 2007.[24] These data are a simple count of the number of approvals granted in any country—if the same GM event has been approved in two countries, it is counted twice. The Big 6 firms accounted for 87 percent of these approvals, with Monsanto claiming nearly half of the total.[25] Outside of the Big 6, the most notable holder of approvals for GM events was Florigene, an Australian company with traits for altering flower color.

In the last column, we use data on the area planted to GM cultivars and event approvals from James (2007), together with other information, to estimate the total "trait-acres" under GM crops and assign them to a given source. A trait-acre consists of a single trait planted on an acre, which means that acreage with stacked genes (e.g., Bt and herbicide tolerance combined in a single cultivar) is counted more than once, depending on the total number of biotechnology traits that are "stacked' in a variety. This effect complicates the estimation, particularly when multiple stacked traits (which may originate from different sources) are widely deployed. We made these assessments for the top four GM crops (soybeans, corn, cotton, and canola) for the 13 countries worldwide in which more than 100,000 hectares of GM crops were planted in 2007.

Some individual trait-crop combinations are easier to assess than others. For example, Roundup Ready soybeans, which account for over 40 percent of all trait-acres, are easy to identify from the data in James (2007). On the other hand, insect-resistance traits are very hard to separate, particularly in corn.[26] To quote an Iowa State University extension publication on integrated pest management, the earliest Bt corn featured "one gene (Cry1Ab), three genetic events (176, Bt-11, MON 810) with event 176 under two trademarks (KnockOut and NatureGard) and events Bt-11 and MON 810 under one trademark (YieldGard)."[27] Furthermore, these Bt events were associated with a number of different companies: Monsanto; Northrup King/Sandoz/Novartis/Syngenta; Ciba-Geigy/Novartis/Syngenta; and Mycogen/Dow.[28]

Based on the level of actual planting of GM crops in 2007, Monsanto traits clearly dominated, with approximately 85 percent of total global trait acres. Four of the other five Big 6 companies had some planted trait acreages, although it is difficult to estimate exact areas. Bayer was notable in that it held the most widely used events for resistance to glufosinate, a herbicide alternative to glyphosate. Glufosinate-tolerant ("Liberty Link") canola varieties were deployed early in Canada and have achieved a minor but notable position in both Canadian and U.S. canola production. Early glufosinate-tolerant events can be traced to Plant Genetic Systems, a biotechnology firm that was acquired first by AgrEvo, then became part of Aventis and eventually Bayer CropScience. Glufosinate tolerance is also being deployed in corn, cotton, and soybeans, but market share up to 2007 appeared to be relatively small for these crops. The only entity outside of the Big 6 companies with notable trait acreage in 2007 was the Chinese Academy of Agricultural Sciences, a public institution, with measureable contributions to Bt cotton planted in both China and India.

Patent data are probably more useful for testing hypotheses on the structure of the seed-biotechnology industry than they are as indictors for the amount of research investment. The amount of patenting in an industry depends on other factors in addition to research spending, such as changes in intellectual property policy (the advent of biotechnology changed and expanded what is considered patentable material in crops). Agricultural biotechnology patent data have been used to test a number of hypotheses about market structure, especially the relationship between intellectual property and mergers, acquisitions, and divestitures. This research has shown that these changes to market structure have been motivated by a need to combine complementary technology assets such as core germplasm, GM traits, and research tools (Graff et al., 2003); the desire to acquire greater depth within a particular technological area (King and Schimmelpfennig, 2005); and the relative enforceability of patents (Kalaitzandonakes and Bjornson, 1997; Marco and Rausser, 2008).

Patent and field trial data can also provide evidence on whether or not new entrants are a significant source of new innovations (Brennan et al., 2005). The USDA/APHIS field trial data in table 2.4 show more concentration than the biotechnology patent data, but Brennan et al. find that concentration in both patents and field trials has increased over time. They claim that new firm entry has increased, but concentration at the top of the industry has also increased. By some measures, U.S. cultivar patents are even more concentrated than other indicators, and at the global level, actually planted GM traits are dominated by a single firm. These concentration indicators may change as other major seed-biotechnology companies seek to commercialize new varieties and traits. Furthermore, tracking traits will become increasingly complex as multiple GM traits from a variety of firms are inserted into individual varieties.

Factors Influencing Future Trends in Private-Sector Seed-Biotechnology R&D

Structural change in the seed-biotechnology industry may have been greater than in any of the other industries covered in this study. Several factors will continue to influence this structure and the level and composition of private-sector investments in seed-biotechnology R&D:

1) *Potential for future market expansion.* Future market expansion will be determined by, among other things, the potential for greater use of improved seed in general and GM crops in particular in developing countries; by changes in consumer attitudes toward genetic engineering, particularly in high-income countries; by the potential for expansion in biotechnology applications to additional crops; and by the development of newer biotechnology applications, for example, tolerance to drought stress.
2) *Industry structure, seed pricing, and seed-biotechnology R&D.* Considerable attention has been focused on the effects of industry structure on seed pricing and potential distributional effects. For example, Stiegert et al. (2010) summarize a number of studies and find that in the United States, own-market concentration can increase seed prices. At the same time, the study finds that if a more integrated system of production of GM seeds by a few large firms reduces development costs, the resulting price effects may reduce or reverse price increases due to market power. Focusing specifically on the impacts of industry structure on R&D, Schimmelpfennig et al. (2004), using field-trial data from the United States, argue that increases in industry concentration had a negative effect on research intensity in agricultural biotechnology. The evidence reported here on actual R&D expenditures for the global seed industry as a whole shows that the net trend in research intensity was strongly positive during the 1990s, and real private-sector research expenditures have continued to grow since then (see table 2.3). Alternatively, Kalaitzandonakes et al. (2010) show that the value of price premiums and markups for GM seed in the U.S. corn and soybean seed industries did not exceed R&D expenditures until 2007, and the authors cite this as evidence of dynamic efficiency in these industries.

3) *Interaction between other policies and seed-biotechnology R&D.* Concerns about potential anti-competitive behavior in the seed-biotechnology sector have expanded beyond distributional effects of seed pricing to questions of whether antitrust policy needs to consider R&D concentration and behavior in trait markets as well as in ultimate seed markets (Moss, 2009; Monsanto, 2009). Regulatory policy as some of the first GM traits (e.g., first generation herbicide tolerance) come off-patent may also determine the potential for entry by firms marketing generic GM crops. Small generic firms that might otherwise be willing and able to develop and market GM seed with an off-patent trait might be less able to meet regulatory compliance requirements than the larger firms that initially marketed the technology when it was still under patent.

REFERENCES

AGROW Reports. 2007. *GM Crops*, Informa UK Ltd., London, July.

AGROW Reports. 2006. *Global Seed Industry*, Informa UK Ltd., London, September.

Brennan, Margaret, Carl Pray, Anwar Naseem, and James F. Oehmke. 2005. "An Innovation Market Approach to Analyzing Impacts of Mergers and Acquisitions in the Plant Biotechnology Industry," *AgBioForum* 8 (2&3): 89-99.

Cahoon, Richard S. 2007. "Licensing Agreements in Agricultural Biotechnology," in A. Krattiger, R.T. Mahoney, L. Nelsen et al. (eds.), *Intellectual Property Management in Health and Agricultural Innovation: A Handbook of Best Practices*, Oxford, UK: MIHR, and Davis, CA: PIPRA, pp. 1009-1016, available at: www.ipHandbook.org.

Cavalieri, Anthony. 2011. Former Vice-President and Director of Trait and Technology Development, Pioneer Hi-Bred International, 2009, personal communication.

Context Network. 2007. "Sizing Up the Seed Market," *Seed World*, pp. 10-12.

Copping, Len. 2003. "The Evolution of Crop Protection Companies," *Pesticide Outlook* 14 (December): 276-78.

Economic Report of the President. 2009. U.S. Government Printing Office.

Enriquez, Juan. 1998. *Gene Research, The Mapping of Life, and the Global Economy*, Harvard Business School Case 599-016, Cambridge, MA.

Falck Zepeda, José, and Joel I. Cohen. 2006. Biosafety Regulation of Genetically Modified Orphan Crops in Developing Countries: A Way Forward," in Richard E. Just, Julian M. Alston, and David Zilberman (eds.), *Regulating Agricultural Biotechnology: Economics and Policy*, New York: Springer.

Fernandez-Cornejo, Jorge. 2004. *The Seed Industry in U.S. Agriculture: An Exploration of Data and Information on Crop Seed Markets, Regulation, Industry Structure, and Research and Development*, Agriculture Information Bulletin Number 786, U.S. Department of Agriculture, Economic Research Service, available at: www.ers.usda.gov/publications/aib786/

Frey, Kenneth J. 1996. *National Plant Breeding Study-I: Human and Financial Resources Devoted to Plant Breeding Research and Development in the United States in 1994*, Special Report 98, Iowa Agricultural and Home Economics Experiment Station.

Fulton, Murray, and Konstantinos Giannakas. 2001. "Agricultural Biotechnology and Industry Structure," *AgBioForum* 4, 2: 137-151.

Graff, Gregory D., Gordon C. Rausser, and Arthur A. Small. 2003. "Agricultural Biotechnology's Complementary Intellectual Assets," *Review of Economics and Statistics* 85, 2 (May, 2003): 349-363.

Heisey, Paul, and David Schimmelpfennig. 2006. "Regulation and the Structure of Biotechnology Industries," in Richard E. Just, Julian M. Alston, and David Zilberman (eds.), *Regulating Agricultural Biotechnology: Economics and Policy*, New York: Springer.

Heisey, Paul W., C.S. Srinivasan, and Colin Thirtle. 2001. *Public Sector Plant Breeding in a Privatizing World*, Agriculture Information Bulletin Number 772, U.S. Department of Agriculture, Economic Research Service, available at www.ers.usda.gov/publications/aib772/

Howard, Philip H. 2009. "Visualizing Consolidation in the Global Seed Industry: 1996-2008," *Sustainability* 1(4): 1266-1287, available at: www.mdpi.com/journal/sustainability/

Huffman, Wallace E., and Robert E. Evenson. 2006. *Science for Agriculture: A Long-Term Perspective, 2nd edition*, Ames, IA: Iowa State University Press.

International Seed Federation. "Estimated Value of the Domestic Seed Market in Selected Countries, International Seed Federation, Seed Statistics," Nyon, Switzerland, Current estimates available at: www.worldseed.org/isf/seed_statistics.html.

James, Clive, 2007. *Global Status of Commercialized Biotech/GM Crops: 2007*, ISAAA Brief No. 37, International Service for the Acquisition of Agri-Biotech Applications, Ithaca, NY.

Just, Richard E. 2006. "Anticompetitive Impacts of Laws That Regulate Commercial Use of Agricultural Biotechnologies in the United States," in Richard E. Just, Julian M. Alston, and David Zilberman (eds.), *Regulating Agricultural Biotechnology: Economics and Policy*, New York: Springer.

Just, Richard E., and Darrell L. Hueth. 1993. "Multimarket Exploitation: The Case of Biotechnology and Chemicals," *American Journal of Agricultural Economics* 75, 4 (November): 936-945.

Kalaitzandonakes, Nicholas, and Bruce Bjornson. 1997. "Vertical and Horizontal Coordination in the Agro-Biotechnology Industry: Evidence and Implications," *Journal of Agricultural and Applied Economics* 29(1): 129-139.

Kalaitzandonakes, Nicholas, Julian M. Alston, and Kent J. Bradford. 2006. "Compliance Costs for Regulatory Approval of New Biotech Crops," in Richard E. Just, Julian M. Alston, and David Zilberman (eds.), *Regulating Agricultural Biotechnology: Economics and Policy*, New York: Springer.

Kalaitzandonakes, Nicholas, Alexandre Magnier, and Douglas Miller. 2010. "A Worrisome Crop? Is There Market Power in the U.S. Seed Industry?" *Regulation* 33(4): 20-26.

King, John, and Paul Heisey. 2003. "Ag Biotech Patents: Who's Doing What?" *Amber Waves*, Vol. 1, Issue 5, U.S. Department of Agriculture, Economic Research Service, pp. 12-13, available at: www.ers.usda.gov/amberwaves/november03/datafeature/

King, John, and David Schimmelpfennig. 2005. "Mergers, Acquisitions, and Stocks of Agricultural Biotechnology Intellectual Property," *AgBioForum* 8 (2&3): 83-88.

Le Buanec, Bernard. 2007. "Evolution of the Seed Industry During the Past Three Decades," *Seed Testing International* (ISTA News Bulletin) No. 134 (October): 6-10.

Magnier, Alexandre, Nicholas Kalaitzandonakes, and Douglas J. Miller. 2010. "Product Life Cycles and Innovation in the U.S. Seed Corn Industry," *International Food and Agribusiness Management Review* 13, 3: 17-36.

Marco, Alan C., and Gordon G. Rausser. 2008. "The Role of Patents Rights in Mergers: Consolidation in Plant Biotechnology," *American Journal of Agricultural Economics* 90, 1 (February 2008): 133-151.

McCouch, Susan R. 2001. "Is Biotechnology an Answer?" in K. Wiebe, N. Ballenger, and P. Pinstrup-Anderson (eds.), *Who Will be Fed in the 21st Century?* International Food Policy Research Institute, pp. 29-40.

Monsanto. 2009. "Observations on Competition in the U.S. Seed Industry," available at www.monsanto.com/newsviews/Pages/monsanto-submission-doj.aspx (accessed October 2010).

Morris, Michael L., Joseph Rusike, and Melinda Smale. 1998. "Maize Seed Industries: A Conceptual Framework," in Michael L. Morris (ed.), *Maize Seed Industries in Developing Countries*, Boulder, CO: Lynne Rienner, pp. 35-56.

Moss, Diana L. 2009. *Transgenic Seed Platforms: Competition Between a Rock and a Hard Place?* American Antitrust Institute White Paper.

National Research Council (NRC). 2010. *The Impact of Genetically Engineered Crops on Farm Sustainability in the United States*, Washington, DC: National Academies Press.

Pray, Carl E., Jikun Huang, Ruifa Hu, Qihuai Wang, Bharat Ramaswami, and Prajakta Bengali. 2006. "Benefits and Costs of Biosafety Regulation in India and China," in Richard E. Just, Julian M. Alston, and David Zilberman (eds.), *Regulating Agricultural Biotechnology: Economics and Policy*, New York: Springer.

Schimmelpfennig, David E., Carl E. Pray, and Margaret F. Brennan. 2004. "The Impact of Seed Industry Concentration on Innovation: A Study of US Biotech Market Leaders," *Agricultural Economics* 30, 2: 157-167.

Shi, Guanming, Jean-Paul Chavas, and Kyle W. Stiegert. 2009. "Pricing of Herbicide-Tolerant Soybean Seeds: A Market Structure Approach," *AgBioForum* 12(3&4): 326-333.

Shi, Guanming, Jean-Paul Chavas, and Kyle Stiegert. 2010a. "An Analysis of the Pricing of Traits in the U.S. Corn Seed Market," *American Journal of Agricultural Economics* 92(5): 1324-1338.

Shi, Guanming, Kyle Stiegert, and Jean-Paul Chavas. 2010b. *An Analysis of Pricing in Horizontal and Vertical Markets: The Case of the U.S. Cottonseed Market*, FSRG Working Paper 2010-5, University of Wisconsin-Madison, Food System Research Group.

Shoemaker, Robbin. 2001. *Economic Issues in Agricultural Biotechnology*, Agricultural Information Bulletin No. 762, U.S. Department of Agriculture, Economic Research Service.

Stiegert, Kyle W., Guanming Shi, and Jean Paul Chavas. 2010. "Innovation, Integration, and the Biotechnology Revolution in U.S. Seed Markets," *Choices* 25, 2: 5-16.

Traxler, Greg. 1999. "Balancing Basic, Genetic Enhancement and Cultivar Development in an Evolving U.S. Plant Germplasm System," *AgBioForum* 2, 1: 43-47.

U.S. Department of Agriculture, Economic Research Service (USDA/ERS). Agricultural Biotechnology Intellectual Property Database, available at: www.ers.usda.gov/data/agbiotechip/

U.S. Department of Commerce, Technology Administration, Bureau of Industry and Security. 2003. *A Survey of the Use of Biotechnology in U.S. Industry*, October.

Virginia Polytechnic Institute and State University. Search USDA Releases and Notifications, Information Systems for Biotechnology, Blacksburg, VA, available at: www.isb.vt.edu/Default.aspx.

End Notes

[1] By "seed," we refer to all planting material used in crop production, including seed grains, cuttings, seedlings, and other plant propagation materials. Our defi nition of the seed "market" also includes transactions for seed traits, including licensing of genetic material used in seed production.

[2] We used Context Network estimates of the value of farmer-saved seed between 2001 and 2006 and extrapolated for earlier and later years. We followed this procedure to maintain consistency in sources for the different categories of seed. However, given substantial areas planted to farmer-saved seed in developing countries, and even seed for some open-pollinated crops in highincome countries, underestimates are likely. More accurate estimates could be found by performing a country-bycountry analysis of seed markets for major crops, an exercise beyond the scope of this study.

[3] Both the estimate of regional seed market shares for Asia-Pacifi c and other estimates, like the share of the global seed market held by rice seed, are particularly infl uenced by how information from China's governmentowned seed enterprises is defi ned and measured.

[4] Although the terms of licensing and royalty agreements are, in general, not public information, most such agreements include statements that define property rights pertaining to each party to the agreement, the cost of the license, the mechanisms for royalty payments, potential philanthropic or humanitarian use, technology stewardship, and enforcement and litigation provisions (Cahoon, 2007). One feature of such agreements that has gained considerable recent attention concerns the rights of the licensee to combine the licensed technology with other technologies, either self-developed or licensed from alternative sources.

[5] Available at www.nass.usda.gov/Publications/Ag_Statistics/index.asp.

[6] Available at www.nass.usda.gov/QuickStats/Create_Federal_All.jsp.

[7] Available at www.ers.usda.gov/data/biotechcrops/.

[8] Available at www.ers.usda.gov/data/arms/app/default.aspx?survey_abb=CROP. We interpolated to calculate seeding rates for years in which they were not estimated.

[9] This method may slightly overstate the germplasm value for self-pollinating crops like soybeans, for which farmers may use part of their own harvest as seed for the next season.

However, in recent years, nearly all soybean seed in the United States has been purchased. Cotton is also selfpollinating, but almost universal use of purchased seed in cotton began much earlier than it did in soybeans.

[10] Kalaitzandonakes et al. (2010) also analyze the pricing of different types of GM traits, either individually or in combination. Much of the data analyzed by Shi et al. (2009, 2010a, 2010b) and by Kalaitzandonakes et al. suggest that component pricing is generally not additive—that is, the price premium for seed containing two or more GM traits is not the sum of the individual premium for the traits taken separately.

[11] For the most part, the data presented by Shi et al. (2009; 2010a; 2010b) indicate slightly higher estimates of the proportion of crop land planted to biotech crops, somewhat higher estimates of the proportion of crop land planted to stacked varieties, and somewhat lower prices for biotech crops than the NASS data.

[12] BASF took a partial interest in Svalöf Weibull, a Swedish seed company, in 1998 but divested in 2008. Svalöf Weibull changed its name to Lantmännen SW Seed in February 2010.

[13] For example, between 2005 and 2010, Dow acquired or reached commercial agreements with nearly 10 regional seed companies, both in the United States and elsewhere, whose primary focus is corn (maize).

[14] Dupont took an initial fi nancial interest in Pioneer in 1997 and took full ownership in 1999.

[15] Estimates of seed market shares for the United States are generally not publically available, although a private source is GfK Kynetec (formerly dmrkynetec). Using GfK Kynetec data, an online Monsanto report traces the corn, soybean, and cotton seed market shares of the largest integrated seed-chemical companies between 1997 and 2009 (see www.monsanto.com/newsviews/Pages/monsanto-submission-doj.aspx). Another consulting company, Context Network, also produces seed market share estimates, and one of its online reports estimates corn seed market shares for leading companies for 2008 (see www.contextnet.com/Focus%20Papers/Seed/Consolidation%20Direction%20Where%20and%20Why%20the%20Seed%20Industry%20is%20Headed%20Sieker%204%2008.pdf).

[16] Generic fi rms are fi rms that do not conduct research on new trait discovery but rather focus on commercializing varieties and traits that have gone off-patent. So far, generic fi rms are virtually nonexistent in the crop seedbiotechnology industry, but they play an important role in the agricultural chemical industry (see chapter 3).

[17] We refer to these companies as "dedicated" agricultural biotechnology companies in that agricultural applications appear to be the main market for their product development efforts. Not included in this category are many other companies that may provide some technology services to agriculture but whose main product focus is in other economic sectors. Names of these companies are available from the authors upon request.

[18] Marker-assisted breeding is an indirect process in which selection for a trait of interest is based on a genetic or other marker associated with the trait, but not the trait itself.

[19] In table 2.3, we include research expenditures by legacy companies that also conducted both agricultural chemical and seed-biotechnology research as part of the Big 6. These include AgrEvo, Astra-Zeneca, Aventis, Ciba-Geigy, Novartis, and Sandoz.

[20] Le Buanec's estimates of biotechnology research as a percentage of total seed-biotech research in 2003 are as follows: Monsanto 80 percent; Dupont/Pioneer 50 percent; Syngenta 60 percent; BASF 100 percent; Dow 85 percent; Bayer Crop Science 85 percent; and Limagrain 10 percent. Applied to our estimates of research expenditures in 2003, this implies an estimate that about 70 percent of all Big 6 seed research was devoted to biotechnology in 2003, with the rest of our estimates based on the assumptions listed in the text. The overall estimate of 50 percent may be roughly the same today, or even higher, as

large seed companies outside of the Big 6 may be increasing biotechnology investments. However, Syngenta has apparently decreased its spending on crop biotechnology while increasing its overall seed research in recent years, if the reported research expenditure devoted to "business development" in its annual reports can be assumed to be biotechnology research. This has been our working assumption for Syngenta. Other companies do not make this distinction in reporting their R&D expenditures.

[21] As mentioned in footnote 17, some privately collected data are available for purchase from such companies as GfK Kynetec. These data focus particularly on large markets such as the United States, on commercial seed sales as reported by farmers, and on major crops such as corn, soybeans, and cotton for which many varieties include GM traits. Magnier et al (2010), for example, use data purchased from GfK Kynetec to develop and analyze a fairly complete list of corn hybrids sold in the United States between 1997 and 2009. These data are also the basis for Shi et al.'s (2009; 2010a; and 2010b) analysis of biotech trait pricing for U.S. corn, soybeans, and cotton.

[22] Patent counts have often been used as an indicator of innovative activity. Huffman and Evenson (2006) use U.S. data on patents issued to private inventors in four agricultural areas as a proxy for private agricultural R&D capital. Brennan et al. (2005) present annual concentration ratios for both agricultural biotechnology patents and agricultural biotechnology field trials in the United States as possible indicators of competitiveness in markets for agricultural biotechnology. Nonetheless, patent data used in this way suffer from some limitations: the propensity to use patents or other forms of intellectual property protection might vary over time, by crop, and over jurisdiction (e.g., the application of utility patents to crop cultivars has been primarily a U.S. phenomenon). The propensity to patent also varies depending on the technology. Research tools, traits for genetic engineering, or GM cultivars, for example, might be more likely to be patented than nonbiotechnology research outputs. Also, intellectual property strategies may vary among companies.

[23] Some of the indicators when traced across time show defi nite patterns. For example, concentration in fi eld trial permits has been consistently high (CR 4s of 60 to 100 percent) (Fernandez-Cornejo, 2004). Concentration ratios (CR 10) for broadly defi ned agricultural biotechnology patents fell over the 1990s if mergers and acquisitions are not considered. If mergers and acquisitions are accounted for, the CR 10 began to rise after 1995 (King and Heisey, 2003).

[24] A GM "event" is an instance where a specifi c gene has been introduced into a particular crop. Subsequent introduction of the same gene into another variety of the same crop is not considered to be a separate event.

[25] If GM event approvals are counted only once (eliminating double counting of the same GM seed approved for release in more than one country), Monsanto's share of total global approvals falls from 49 to 38 percent, since that Monsanto has been particularly active in obtaining multiple approvals for the same GM events. The two most widely approved GM events, GTS 40-3-2 (glyphosate-tolerant or "Roundup Ready" soybeans) and MON 810 (Bt corn expressing the Cry1Ab protein) both belong to Monsanto.

[26] In addition, event approvals are sometimes issued jointly to more than one company.

[27] See www.ipm.iastate.edu/ipm/icm/node/946/print.

[28] A number of different Bt genes are available, and they are now aimed at the control of different pests (e.g., the European corn borer and rootworm), contributing to the complexity of the situation with respect to insect resistance.

Index

#

20th century, viii, 115

A

access, 52, 128, 130, 132
accounting, 14, 120
acid, 13, 14, 70, 85
acquisitions, 81, 126, 127, 128, 130, 131, 133, 136, 137, 141, 149
active compound, 71
Africa, 14, 118, 134
agencies, 71, 111
Agricultural Research Service (ARS), 3, 68, 76, 104
agriculture, viii, 36, 76, 81, 87, 102, 115, 130, 131, 137, 139, 148
air quality, 44
alfalfa, 3, 5, 12, 13, 16, 84, 122
amino, 13
amino acid, 13
amylase, 70
analytical framework, 99
anti-cancer, 13
antioxidant, 13
antitrust, 127, 143
Argentina, 14, 88
ARS, 5, 10, 68
Asia, 118, 134, 147
ASS, 17, 35
assessment, 64
assets, 67, 141
assumptions, 99
atmosphere, 42
attitudes, 5, 78, 142
audit, 52
Australia, 98
authors, 76
availability, 79, 97

B

bacterium, 5, 89
bankruptcy, 131
barley, 84, 98
barriers, 130
base, 30
basic research, 131
beef, 47, 99
beer, 48, 98
behavior, 108
Beijing, 63, 96, 97, 107
Belgium, 47, 97
benchmarks, 6
beneficiaries, 99
benefits, vii, viii, 1, 6, 14, 31, 34, 36, 37, 40, 42, 49, 54, 72, 73, 75, 76, 78, 79, 93, 96, 97, 99, 102
benign, 42, 43

beta-carotene, 13, 14, 85, 86
bias, 34, 45, 71, 72, 113
biodiversity, 56
biological sciences, 81
biotech firms, vii, 1, 2, 6, 78, 99
biotic, 71
Brazil, 14, 49, 50, 60, 88
breeding, viii, 12, 70, 72, 79, 80, 81, 84, 115, 116, 119, 120, 123, 124, 132, 135, 148
business model, 126
businesses, 81, 126

C

calorie, 13
Canada, 88, 98, 103, 112
cancer, 13
carbohydrate, 13, 14, 85
carbon, 36, 58
Caribbean, 118
carotene, 13, 14, 85, 86
category a, 119, 123, 148
causal relationship, 37
CBS, 103
cell biology, 130
chemical(s), 14, 21, 30, 36, 71, 72, 88, 90, 94, 112, 121, 122, 123, 126, 127, 129, 131, 133, 134, 148
chemical industry, 148
Chicago, 57, 104
chicken, 47
China, 14, 35, 48, 49, 50, 63, 80, 88, 96, 97, 98, 107, 119, 141, 146, 147
cholesterol, 49, 50
classes, 119, 136
cleaning, 4, 43, 73
close relationships, 126
cluster analysis, 49
coffee, 13, 14
collaboration, 69, 128, 132
color, 13, 14, 70, 85, 97, 140
commerce, 56
commercial, vii, 1, 6, 7, 13, 14, 18, 21, 39, 43, 53, 71, 72, 79, 80, 81, 85, 89, 90, 116, 117, 118, 121, 128, 129, 148, 149
commodity, 17, 99, 118
communities, 58
comparative advantage, 91
competitiveness, 26, 149
complement, 30
complementarity, 28, 79, 126
complexity, 149
compliance, 39, 42, 73, 130, 135, 143
composition, 43, 65, 142
compounds, 49, 50, 71
concentration ratios, 128, 129, 149
confidence, 102
Consensus, 111
conservation, 16, 30, 36, 37, 38, 44, 72, 78, 94, 95
consulting, 148
consumer choice, 99
consumer concerns, viii, 75, 78, 79
consumer surplus, 113
consumers, vii, viii, 2, 4, 5, 6, 44, 45, 46, 47, 48, 49, 50, 51, 52, 73, 76, 77, 78, 79, 95, 96, 97, 98, 99, 102, 113
consumption, 116
control, 77, 80, 84, 85, 88, 89, 90, 93, 113
control group, 23
Cooperative State Research, Education, and Extension Service, 76
corn, 75, 76, 77, 79, 80, 81, 83, 84, 88, 89, 90, 91, 92, 93, 94, 112, 113
correlation, 72
cost(s), 26, 28, 30, 40, 41, 44, 73, 75, 79, 81, 85, 88, 89, 91, 99, 113, 119, 130, 132, 147
cost saving, 30
costs of production, 49
covering, 36
CRM, 104
crop producers, 4, 39, 43, 44
crop production, viii, 54, 59, 115, 147
crop residue, 36, 72, 113
CT, 62

cultivars, 18, 79, 90, 112, 136, 137, 138, 139, 140, 149
cultivation, 54

D

data availability, 135
data collection, 17
database, 10, 53, 78, 137, 138, 139
decisions, 113
degradation, 36, 94
Delta, 83
Denmark, 122
Department of Agriculture, 7, 16, 25, 26, 58, 60, 61, 62, 68, 69, 75, 81, 104, 105, 109, 110, 111
depth, 141
deregulation, 3, 13, 70, 87
detection, 9, 70
developing countries, 4, 44, 49, 52, 116, 131, 142, 147
distribution, 26, 28, 57, 99, 102, 119, 129, 130
diversity, vii, 1, 4, 39, 43, 102
DNA, viii, 10, 70, 112, 115, 130, 131
domestication, 129
dominance, 116, 136
double counting, 149
drought, 2, 3, 5, 10, 13, 14, 16, 77, 79, 83, 85, 86, 142
drugs, 71

E

eating, 79
economic incentives, 42, 99
economic theory, 40
economics, 40, 54, 59
economies of scale, 81
editors, 108
Education, 76
elasticity, 99, 103
election, 137, 139
employment, 31, 93
endogeneity, 72
enforcement, 147
engineering, 5, 44, 70, 79
environment, 37, 40, 52, 56, 70, 78, 81, 87, 94, 95, 136
environmental impact, vii, 2, 77, 94
Environmental Protection Agency, 7, 39, 60, 69, 81, 111
environmental quality, 36
environmental stress(s), 5, 79
EPA, 7, 8, 39, 42, 69, 81, 102
epidemic, 13, 85
equilibrium, 113
equipment, 4, 42, 43, 73
erosion, 36, 94
ERS, 5, 6, 16, 17, 23, 25, 26, 30, 31, 32, 36, 53, 54, 69, 71, 73, 76, 78, 91, 93, 94, 99, 113, 122, 137, 138, 139, 147
EU, 4, 45, 49, 52, 53, 106
Euro, 98
Europe, viii, 46, 52, 54, 75, 79, 108, 118, 120, 128, 134
European Commission, 62
European Community, 57
European consumers, viii, 47, 76, 78
European Union, 44, 78, 95, 99, 106, 107
evidence, 3, 4, 23, 30, 31, 42, 53, 64, 73, 97, 141, 142
evolution, vii, 1, 2, 4, 38, 39, 40, 41, 43, 44, 54, 73
exercise, 147
expenditures, 6, 17, 80, 112, 132, 133, 134, 135, 142, 148
exploitation, 102
exposure, 14, 73, 88, 94

F

farms, 4, 31, 32, 33, 34, 42, 43, 44, 73
fat, 99
FDA, 7, 8, 58, 71, 79, 81, 96
Federal Register, 8, 58, 69, 71, 81, 109, 112
fiber, 5, 13, 14, 79, 85
field crops, 12, 64, 80, 83, 84, 94, 118, 121, 123

field tests, vii, 1, 7, 21, 31, 81, 91
field trials, 137, 141, 149
filters, 71
financial, 71, 91
financial performance, 71
firms, 75, 78, 81, 99, 112, 113
first generation, vii, 5, 49, 75, 76, 79, 143
flavor, 3, 7, 10, 13, 14, 71, 83, 85, 89
flexibility, 30, 31, 78
flowers, 13, 14, 85
food, 4, 5, 8, 46, 47, 48, 49, 50, 51, 52, 53, 54, 73, 76, 78, 79, 81, 95, 96, 97, 98, 99, 112, 113, 116, 137, 139
Food and Drug Administration, 7, 58, 71, 79, 81, 96, 112
food production, 96, 139
food products, 48, 52, 76, 78, 99
Ford, 67, 110
France, 47, 48, 98, 122, 123
frost, 13, 14, 71, 85
fruits, 13, 14, 85
fungi, 12

G

GE crops, vii, viii, 1, 2, 3, 5, 6, 14, 15, 16, 18, 28, 31, 32, 54, 70, 71, 75, 77, 78, 79, 88, 89, 90, 91, 93, 94, 99, 113
GE seeds, vii, 1, 3, 13, 14, 18, 21, 28, 53, 54, 95
gene(s), 7, 9, 18, 51, 70, 71, 73, 89, 90, 112, 130, 140, 149
gene expression, 8
generation, 79, 113
genetic engineering, 44, 70, 79, 112, 130, 137, 139, 142, 149
genetics, viii, 28, 112, 115, 128, 129, 133
genomics, 137, 139
geography, 4
Georgia, 68
Germany, 45, 47, 50, 98, 122, 123
global scale, 14
globalization, 134
glyphosate, vii, 1, 3, 4, 5, 25, 28, 30, 36, 39, 42, 43, 44, 54, 55, 57, 59, 64, 66, 72, 73, 81, 94, 127, 140, 149
goods and services, 97
grass, 30, 118, 123
Gross Domestic Product, 117, 134
groundwater, 41
groups, 113
growth, 25, 26, 31, 128, 129
guidance, 69
guidelines, 44, 112

H

harvesting, 13
Hawaii, 13, 85
hazards, 71
health, 36, 40, 52, 72, 122, 128, 129, 133
health effects, 40, 72
health risks, 36
heat, 99
herbicide, vii, 1, 2, 5, 7, 10, 14, 15, 16, 18, 24, 25, 26, 27, 28, 29, 30, 31, 33, 35, 36, 37, 38, 39, 42, 43, 54, 55, 57, 66, 67, 72, 77, 78, 79, 81, 83, 87, 88, 90, 101, 112, 126, 127, 130, 140, 143
herbicide-tolerant crops, vii, 1, 30, 37, 38
heredity, 70
host, 73
House, 107, 110
household income, 3, 31, 78, 93
human, 64, 71
hybrid, 25, 80, 90, 119
hybridization, 119

I

ideal, 41
identification, 131
identity, 54
improvements, 21
incentives, 77
incidence, 72

income, 3, 31, 44, 45, 49, 78, 93, 116, 120, 142
India, 14, 49, 50, 54, 63, 141, 146
indication, 86, 96
individuals, 40, 137
industry(s), viii, 2, 6, 28, 51, 76, 80, 81, 115, 116, 121, 126, 127, 128, 129, 130, 131, 133, 134, 135, 141, 142, 148
infancy, 43
infestations, 16, 21, 72, 77, 93
inflation, 4, 26, 80, 117, 134
ingredients, viii, 4, 44, 47, 48, 52, 53, 73, 76, 78, 79, 95, 96, 97, 99, 112
insecticide, 3, 31, 32, 34, 35, 72, 73, 90, 94, 112, 113
insect-resistant crops, vii, 1, 77
insects, 3, 5, 6, 16, 21, 32, 35, 39, 40, 89
insight, 97
institutions, 3, 10, 131, 136, 137, 139
instruments, 102
intellectual property, viii, 77, 80, 115, 130, 141, 149
intellectual property rights, 77, 80, 130
international trade, 102
inventors, 149
invertebrates, 7, 81
investment(s), 28, 80, 81, 112, 123, 127, 131, 133, 134, 135, 136, 141, 142, 149
investors, 131
Iowa, 34, 42, 61, 140, 144
IPR, 80
iron, 13, 14, 85, 86
irrigation, 71
isolation, 34
issues, vii, 2, 7, 8, 73, 77, 81
Italy, 50, 55, 98, 103

J

Japan, 47, 80, 98, 99, 122
joint ventures, 81, 128
Jordan, 66, 67
jurisdiction, 149

K

Kenya, 49, 51, 62

L

labeling, 73, 78, 99
labor, 89
larvae, 42
Latin America, 118, 134
laws, 116
leaching, 41
lead, 31, 32, 72, 116
legal protection, 80
leisure, 31
Lepidoptera, 55, 67, 103, 110
life sciences, 112
lignin, 13, 85
litigation, 147
livestock, 71
Louisiana, 77
lower prices, 148
lycopene, 13
lysine, 70

M

machinery, 128, 129
magnitude, 45
majority, 23, 53, 91, 94, 139
mammals, 72
management, vii, viii, 1, 2, 3, 4, 14, 16, 23, 28, 30, 31, 39, 42, 43, 54, 71, 72, 73, 75, 78, 79, 86, 88, 89, 93, 126, 140
market, 78, 79, 80, 83, 84, 85, 89, 91, 95, 97, 99, 113
market concentration, 80, 142
market position, 136
market share, 83, 91, 113, 123, 129, 141, 147, 148
market structure, 141
marketing, viii, 13, 26, 44, 53, 73, 95, 99, 115, 126, 130, 143
marketing strategy, 99

marketplace, 131
markets, 99, 106, 113
Maryland, 57, 104
materials, 147
measures, 113
meat, 52, 53
media, 67, 73, 132
medicine, 62
melon, 12, 84
mergers, 81, 126, 130, 133, 141, 149
methodology, 45
Mexico, 41
microbial communities, 58
microorganisms, 7, 79, 81
Middle East, 118, 134
mineralization, 58, 65
model(s), 45, 113
modifications, 7
monopoly, 113
monopoly profits, 113
Moon, 45, 47, 61, 65, 98, 108
movement, 78
Multilateral, 111
multinational companies, 119
multinational corporations, 102, 121

N

NAFTA, 118, 134
National Academy of Sciences, 56, 81
National Research Council, 13, 18, 32, 65, 108, 109, 146
National Survey, 64
natural hazards, 71
Natural Resources Conservation Service (NRCS), 69
Netherlands, 108, 122
neutral, 97
New Zealand, 45, 47
nicotine, 70, 99
nitrogen, 13, 14, 85, 86
North America, 118, 123, 128
Norway, 47, 48, 60, 98, 106
NRC, 12, 23, 26, 28, 31, 36, 37, 39, 43, 65, 71, 81, 129, 146
NRCS, 36, 44, 69
nutrient, 5, 71, 79
nutrition, 3, 4, 7, 10, 44, 49, 83, 96

O

oil, 13, 16, 46, 47, 48, 49, 50, 70, 85, 86, 97, 98, 99
oils, 76, 79, 85
operating costs, 17
operations, 3, 112, 139
operator, 89
opportunities, 126
order, 81
organic food, 99, 113
organism, 8, 10, 13, 81, 137, 139
oversight, 7, 8, 13, 81
ownership, 126, 135, 148

P

Pacific, 118, 134
Pakistan, 14
Paraguay, 14
parents, 73
participants, 45, 129
patents, 70, 80, 136, 137, 138, 139, 141, 149
peer review, 81
permission, 87
permit, 7, 8, 10, 71, 78, 79, 81
personal communication, 143
pest populations, 41, 73
pesticide, viii, 2, 8, 16, 18, 19, 21, 30, 31, 38, 40, 41, 42, 50, 54, 56, 71, 72, 75, 76, 78, 79, 81, 88, 89, 90, 91, 93, 94, 96, 113
pests, 3, 18, 21, 28, 35, 38, 39, 40, 41, 49, 71, 89, 90, 149
pharmaceutical(s), 5, 14, 71, 79, 81, 84, 86, 112, 126
phenotype, 11, 71
pipeline, 65, 86
plant breeding, viii, 70, 79, 80, 81, 115, 116
plants, 7, 41, 71, 72, 73, 79, 81, 131, 137
platform, 130

policy, 67, 73, 141, 143
pollen, 42, 43, 44
pollution, 40
polymers, 127
population, 17, 34, 39, 40, 41
population density, 34
potato, 12, 51, 71, 85, 89
potential benefits, vii, 1, 6, 76, 102
poultry, 68
premiums, 91, 99, 113
present value, 40
preservation, 102
President, 109, 117, 134, 143
pressure, 92, 93
price deflator, 117, 134
price effect, 142
price index, 26
prices, 93, 96, 99, 113
pricing behavior, 26
principles, 64
private firms, 81, 137
private sector, 26, 116, 131, 133
probability, 17, 23, 24, 30, 31, 72
producers, viii, 4, 23, 27, 30, 39, 41, 43, 44, 53, 75, 77, 93, 99
production, 79, 81, 99
production costs, 55, 73, 99
productivity, 76, 79
profit, 26, 71
profit margin, 26
profitability, 21, 23, 30, 89, 93, 135
propagation, 147
property rights, 6, 77, 80, 147
protection, 73, 77, 80, 93, 116, 123, 126, 127, 134, 136, 149
public expenditures, 6, 80
public sector, 26, 116, 117, 128
PVP, 80

R

rainfall, 71
range, 77, 93
real terms, 4, 26
recombinant DNA, 112
recombination, 70
recommendations, 81
region, 93
regulation(s), 7, 8, 39, 73, 81
relaxation, 41
repair, 130
replication, 130
reproduction, 70
requirements, 4, 39, 41, 42, 118, 143
research institutions, 131
researchers, 13, 19, 20, 23, 30, 31, 32, 36, 39, 52, 73, 85, 91, 92, 96, 99
residuals, 72
residues, 40, 41, 72
resistance, vii, 1, 2, 3, 4, 5, 7, 10, 12, 13, 14, 15, 16, 21, 25, 27, 34, 36, 38, 39, 40, 41, 42, 43, 44, 54, 73, 77, 79, 83, 84, 85, 86, 87, 88, 90, 112, 126, 140, 149
resources, 41, 62, 135
response, 44, 127
responsiveness, 24, 52, 72, 113
restrictions, 41
retail, 4, 53
retention, 94
returns, 77, 80, 91, 92, 93, 112, 113
revenue, 17, 119
rewards, 34
rice, 84, 97, 98
rights, 6, 77, 80, 136, 147
risk(s), vii, 1, 6, 7, 8, 13, 36, 49, 72, 76, 81, 87, 93, 97, 102, 131
Romania, 47, 49, 57
root, 42
royalty, 147
rules, viii, 115
runoff, 36, 72, 94

S

safety, 9, 31, 70, 72
sales, 81, 97, 113
salinity, 13, 14, 85, 86
savings, viii, 14, 21, 30, 31, 54, 59, 75, 79, 88, 90, 113
science, 126

scope, 28, 54, 76, 78, 126, 147
second generation, 49
Secretary of Agriculture, 69
seed industry, viii, 2, 6, 28, 76, 80, 81, 115, 116, 126, 129, 130, 131, 133, 134, 135, 142
seeding, 120, 147
seedlings, 147
selectivity, 71, 113
services, 25, 97, 148
shares, 89, 101
shelf life, 13, 99
showing, 89, 136
shrimp, 99
signals, 14
social welfare, 40, 41
soil, 78, 89, 94, 95, 113
soil erosion, 36, 94
South Africa, 14
soybean, 75, 77, 81, 89, 91, 93, 99, 103
Spain, 64
species, vii, 1, 4, 39, 41, 42, 43, 51, 73, 129
spending, viii, 80, 115, 133, 134, 141, 149
Spring, 56, 57
stakeholders, vii, viii, 2, 6, 49, 51, 75, 77, 78, 113
standards, 81
starch, 13, 14, 85, 86
states, 72
statistics, 144
statutes, 7, 81
stock, 40
strategies, 99
stress, 5, 71, 142
structural changes, 127
structure, 81, 129, 130, 141, 142
students, 98
subsidy, 40
substitution, vii, 1, 30, 36, 95
sugar, 84
sugar beet, 12, 84, 123
sugarbeets, 3, 5, 16
sulfonylurea, 66
suppliers, vii, 2, 6, 52, 77, 78, 79, 99
supply, 99, 103, 113
suppression, 32, 34, 56
survival, 38, 40
susceptibility, 41
sustainability, 70, 144
Sweden, 45, 47
Switzerland, 111, 122, 144

T

tactics, 73
Taiwan, 47, 98
Taiwanese students, 48, 98
target, 10, 28, 39, 41, 42, 70, 73
techniques, 37, 70, 72, 79, 99, 112, 113, 116, 130, 137, 139
technology(s), vii, viii, 2, 6, 7, 8, 10, 11, 13, 14, 18, 29, 31, 37, 39, 43, 44, 52, 55, 71, 75, 77, 78, 79, 88, 91, 99, 102, 113, 118, 119, 123, 126, 130, 141, 143, 147, 148, 149
testing, 2, 5, 6, 7, 8, 10, 11, 13, 14, 77, 78, 79, 81, 83, 84, 86, 87, 141
texture, 13, 14, 85
time periods, 41, 53
tobacco, 3, 12, 13, 16, 41, 70, 84, 85, 89
total product, 126
toxicity, 32, 36, 64, 67, 72
toxin, 39, 41, 42
trade, 102
trademarks, 140
transactions, 147
transformation, 80, 81
transmission, 4
transparency, 73
transport, 136, 139
treatment, 23, 72
trees, 84, 85
trial, 71, 141, 142, 149
tumors, 72
turnover, 131

U

U.S. Department of Agriculture, 1, 7, 16, 25, 26, 58, 60, 61, 62, 68, 69, 75, 81, 104, 105, 109, 110, 111, 115, 144, 145, 146, 147
U.S. Department of Commerce, 147
U.S. economy, 80
uncertainty, 90
United Kingdom (UK), 45, 46, 47, 53, 56, 64, 65, 97, 98, 103, 108, 143
United Nations, 59, 70, 111
universities, 25
USA, 58, 105

V

Valuation, 45, 63, 65, 70, 98, 108, 112
variable costs, 71, 92, 113
variables, 22, 72
vector, 10
vegetable oil, 47, 49, 98
vegetables, 12, 13, 14, 50, 84, 85, 123
venture capital, 131
Vermeer, 97, 107
Vietnam, 54
vitamin A, 50
vitamin C, 97
vitamins, 14, 49, 50, 51, 86

W

wages, 71
Washington, 64, 65, 108, 109, 111, 146
water, 36, 40, 44, 71, 72, 94
websites, 53, 124, 132
welfare, 40, 41, 99
wheat, 80, 83, 84
White Paper, 146
wind, 94
Wisconsin, 34, 67, 146
World Health Organization (WHO), 36, 70
worldwide, 88, 118, 119, 120, 136, 140

Y

yield, 3, 4, 14, 18, 21, 24, 34, 49, 56, 71, 85, 88, 89, 90, 91, 93, 112, 113